Intimacy at Work

Intimacy at Work:

How Digital Media Bring Private Life to the Workplace

Stefana Broadbent

Walnut Creek, California

LEFT COAST PRESS, INC.
1630 North Main Street, #400
Walnut Creek, CA 94596
www.LCoastPress.com

ISBN 978-1-62958-094-4 hardback
ISBN 978-1-62958-095-1 paperback
ISBN 978-1-62958-096-8 institutional eBook
ISBN 978-1-62958-097-5 consumer eBook

Library of Congress Cataloging-in-Publication Data:

Broadbent, Stefana.
[L'intimité au travail. English]
Intimacy at work : how digital media bring private life to the workplace / Stefana Broadbent.
pages cm.—(Anthropology and business ; Volume 2)
Includes bibliographical references and index.
ISBN 978-1-62958-094-4 (hardback)
ISBN 978-1-62958-095-1 (paperback)
ISBN 978-1-62958-096-8 (institutional eBook)
ISBN 978-1-62958-097-5 (consumer eBook)
1. Personal Internet use in the workplace. 2. Cell phones—Social aspects. 3. Information technology—Social aspects. 4. Social media. 5. Work environment. 6. Work-life balance. 7. Right to privacy. I. Title.
HF5549.5.P39B7613 2015
302.3'5—dc23

2015026293

Printed in the United States of America

∞™ The paper used in this publication meets the minimum requirements of American National Standard for Information Sciences—Permanence of Paper for Printed Library Materials, ANSI/NISO Z39.48–1992.

Contents

Acknowledgments

This book is about very personal emotions. It tells the stories of people who have been using technology to express their feelings and empathise with people around them. I am infinitely grateful to everyone who has been so generous as to spend time with me reflecting and explaining their intimate thoughts. I know I have not been able to do justice to the richness and depth of the relationships they have recounted.

I also wish to thank my students at UCL in London and the ENSAD in Paris who have helped me with precious interviews and observations. I am especially grateful to Maria Ramirez Angel with whom we shared findings and interviews for many years. My colleagues at Swisscom, Valerie Bauwens, Petra Hutter, Caroline Hirst, Daniel Boos, were indefatigable research partners.

I am indebted to many people for my intellectual meanderings. Francesco Cara, Dan Sperber, and Ed Hutchins have an enduring influence on my understanding of social and cognitive phenomena. Eliana Adler Segre taught me the importance of theories of emotions. My colleagues at UCL Haidy Gesimar, and Daniel Miller have introduced me to the mysteries of material culture. Claire Lobet Maris and Nicole Dewandre have helped me understand the impact of attention. Geoff Mulgan and Stian Westlake at Nesta are opening my eyes to the world of governance and policy.

Finally I wish to thank Jack Meinhardt and the editors at Left Coast Press, Inc., for their patience and support and Timothy de Waal Malefyt for his immediate interest in the book.

Introduction

In the last fifteen years 5 billion people have gleefully adopted the opportunity to stay in continuous contact with people they love.

Their days are now dotted with small interactions with family, partners, and friends through texts, pictures, likes, e-mails, and phone calls. Close scrutiny of what people actually do with all the channels they have at their disposal shows an intensification of exchanges with few close ties, often less than five, that lead to the strengthening of these relationships. All the data on mobile and Internet communication we have at our disposal shows that users in all countries are having short and frequent exchanges with the people who are closest to them. This constant and ubiquitous link between individuals and their loved ones is very emotionally intense, and the feeling of being always within reach provides a profound sense of safety and comfort. Both the staggering speed of adoption (it took 120 years to get 1.2 billion subscribers to use fixed phones, whereas it took only from 2002 to 2008 to increase the number of mobile phones from 1 billion to 4.1 billion), and the universality of the communication practices that have emerged indicate the depth of the emotional response to these channels. However, these billions of short conversations, messages, and endearments are seriously subverting people's relations to the key institutions they inhabit, such as their workplaces or schools. Environments where individuals were expected to be isolated from their private social sphere, such as offices, factory floors, hospitals, or classrooms, are now accessible to personal exchanges. These private conversations are challenging a number of well-rooted conceptions on the need to remove people from their personal social environment in order for them to be productive and effective.

Most organisations for the last 150 years had, in fact, banned the private sphere from their premises, as the majority of Western society has functioned on the principle that attention, isolation, and productivity are strictly interrelated. Although some form of contact from the workplace was possible after the introduction of the telephone in offices in the late 1930s, social factors such as status, power, and trust meant that only the highest professional ranks were allowed to do so.

Intimacy at Work: How Digital Media Bring Private Life to the Workplace, by Stefana Broadbent, 9–14.

Only managers and directors had their own external line. The generalisation of personal digital devices, PCs first and then mobile phones, has meant that private communication has now become possible at all levels of the professional and educational hierarchy, blue-collar workers and managers alike. Not surprisingly, most institutions initially set limits and regulations on where, when, and how much people were allowed to use their private channels. Educational institutions systematically introduced measures to limit mobile device use; administrations initially blocked access to social networking websites; and manufacturing, retail, and transport companies prohibited the use of phones during working hours, with sanctions leading to dismissal in the most serious cases. Restrictions emerged and still are the tightest in the settings where people are trusted the least. In many environments where control is exerted, it is simply assumed that without restrictions, people will abuse the access and be incapable of controlling their attention and their commitment.

What the exceptionally rapid adoption of personal communication devices has shown is that the division of the private and professional realms is arbitrary and often unwelcome. When given the possibility, most people seem to want to be able to keep in touch with their personal social sphere whenever they feel the need for it. Paradoxically, in the last century, while we developed a culture that elevated the family and intimate life to the main social space of individuals, a space that is supposed to provide all the comfort, sustenance, and happiness that the public realm cannot offer, people were made to work and learn in environments that cut them off from their closest ties. From an emotional perspective, therefore, people are living in a society that overemphasises the role of intimate relationships and throws onto the family the ultimate responsibility of caring and nurturing the individual, while these same people are also spending a large portion of their day severed from those relationships. In light of the current profound transformation of the labour market, which is leading to a greater fragmentation and instability of work, with longer periods of under- or unemployment, the role of strong personal ties is ever more prominent. It is not surprising, then, that the moment a channel of communication emerged that could be used to reconnect the personal and professional realms, it was immediately adopted.

Digital communication services have, therefore, proven to be revolutionary in two senses: they are allowing people to maintain a personal space within institutions that banned the private sphere from their premises, and they are offering this possibility to people who by status or condition were particularly unlikely to be granted such a privilege. Current debates about attention and the loss of it must, therefore, be read in light of such a transformation. The fear that access to the Internet will lead to a loss of focus and, ultimately, of productivity is to be analysed in sociocultural terms more than cognitive ones, as the issues they raise have more to do with domination and control than with memory, cognitive abilities, or vigilance.

Approach

The phenomenon we are witnessing, this massive and ubiquitous use of digital communication to stay in touch with a few important relationships, raises many questions, as it involves various levels of human experience, from intimate emotions to social norms, from cognitive processes to institutional regulations and sanctions. Thus, in order to analyze and understand this variety of experiences, I have invoked explanatory frameworks from different disciplines in the social sciences.

The emotional response that we observe when people have intense exchanges with people they love seems to need the explicatory framework of clinical psychology. I have found it impossible to understand the urgency and the need that individuals feel to be able to maintain a permanent sense of connectedness with a few very significant relationships without invoking mechanisms of attachment and separation. In particular many discussions we have had with cell phone users have centred around issues of safety and feeling secure, and this has led us to look into the literature on emotional development.

Digital channels are also being used as media that enact and enable people to perform certain relationships. The profound entrenchment of these media in the communicational patterns of users have made them into cultural artefacts that make relationships manifest and tangible. Friendship, parenting, and loving are enacted in the media and participate in supporting new subjectivities. Theories of material culture provide a background with which to understand how identity and relationships are articulated and assembled around artefacts and symbolic systems.

Conversely, there are also rapid social changes in the way social groups, organisations, and institutions are incorporating these tools and embedding new communication practices in their system of norms and practices (Kaplan 2010). These norms are only understandable in historic terms and within a sociological framework that captures the evolving relation between the public, the professional, and the private spheres. The issues around the role of personal communication in the workplace reflect a tension between private and public that is a debate that has been embraced by sociologists for many years. Here the questions concern the role of the individual in an organisation and the structures of power. A more historical perspective is also needed to understand the origin of the norms that the new practices challenge.

Finally, the tools and services themselves evolve constantly and entrench in their functions cognitive requirements that will determine how they are used and how they will be embedded in social environments. The third theoretical stream is the cognitive one and, in particular, human factors approaches to tasks and accidents. Because one of the main arguments against the introduction of personal communication devices in the workplace is the risks associated with diminished attention, we have to draw on the human factors literature to understand the role of cognitive processes in accidents. Human factors analyses tend to dispel the myth that individual distraction can be the sole culprit in events that involve operators in complex environments.

The Data

The data I present also come from different sources: some are from large surveys and studies done by different research bodies, universities, research foundations, and market research companies. Most of my data come from Europe and the United States, although I have tried, where possible, to mention research done in Western Africa, Colombia, China, and Australia. The phenomena we are observing are so widespread and are happening at such a scale that data cannot be easily collected by any single organisation, and although international bodies such as the ITU (International Telecommunication Union) or some international market research agencies can capture the main parameters of infrastructural diffusion, they cannot usually study specific user behaviours. Conversely, studies that capture the actual usage of different services and details about who communicates with whom and where from are much more rare and require us to delve into myriad more local projects. My quantitative sources, therefore, can also be university studies, in particular regions, as market research studies done by companies involved in the communication sector. In any case, because of the scale of the transformations we are witnessing, it is imperative to combine sources of information and, where possible, have some idea of the extent of the phenomena we are discussing.

Most of my insights, however, come from twenty years of my own ethnographic research on the usage of communication technology at work and at home. I know of no other discipline that has such a naturalistic approach, and I have embraced it as the only way to have a sufficiently holistic view of what is happening. The ethnographic approach has allowed me to systematically gather information not only on people's digital life but also their relationships, their physical and social environments, and the organisation of their time. I will illustrate many points with the narratives of people that my students, colleagues, and I have interviewed over the years. I have tried as much as I could to uphold our participants' voices and stories while protecting their anonymity.

All of my ethnographic research in the home has been done in Europe, mostly in France, Italy, Switzerland, and the UK. Many of the discussions about private communication from the workplace were carried out at home or outside the company premises, as we found it very difficult to be able to have open and frank discussions about this theme at work. Our attempts to interview human resource personnel on these matters were also fraught with difficulty, as it is often considered a delicate subject, about which regulations are not clear-cut. Since 1990, however, I have conducted observations in a variety of work environments in Europe and occasionally in Africa and the United States, and I am familiar with the organisational beliefs that underlie some of the norms that have developed regarding private communication. Lastly, I have spent a certain number of years involved in projects on operator decision making and monitoring processes, what can generically be labelled human factors projects. The current preoccupation with the cognitive impact of personal devices inscribes itself in the long tradition of research on attention, monitoring activities, and accidents. Thus, I have found it useful to reconsider some of the accidents ascribed to mobile phone usage, in the light of what we know in human factors on operators' attention cycles.

Chapters

There is a huge interest in understanding what people are really doing with new technologies in general and communication technology in particular. The topic seems to be an endless object of fascination for media audiences, mostly because people like to be reassured that they are not too different from everyone else. However, the media are by and large supporting a technology hype that systematically misrepresents what average users are doing. Many years of research in digital life provide a more realistic picture of what type of communication is actually going on. Thus, my first two chapters describe how each communication channel contributes to maintain intense and intimate relationships with a close sphere of connections.

The main sections of the book—Chapters 3, 4, and 5—examine the nature of the social transformation at work, especially in the light of new expectations of flexibility. The experience of migration, seasonal employment, and work in general are now considerably different because of the possibility of regular and lengthy contact. Even in less extreme cases, where the separation is just for a few hours each day, exchanges with close ties from workplace or school are modifying the relation to these institutions. All sections of society are finding it possible to bring their personal, intimate social sphere with them into these environments, a privilege that was granted to very few in the past. This intrusion of privacy is extremely subversive because it breaks the equation between isolation and productivity and reduces the potential of social control that comes from isolating individuals from their support network. It also corresponds to a new situation in which people, as professional lives become more precarious and fragmented, have to rely on their most personal ties for ensuring long-term economic stability. Chapter 3 will describe and provide data on this new intimacy and the relation to the changing work environment. Chapter 4 will point out some of the social reasons why laws and regulations are emerging to restrict the use of personal communication devices and channels. I will argue that they are directed towards professional and social sectors that have traditionally been more controlled, such as children, women, and blue-collar workers.

In Chapter 5 I will demonstrate how accidents involving mobile phones—by transport personnel, for instance—can help us understand what is happening at work. Cognitive explanations of attention cycles can far better explain how people are using texting or e-mailing not only to handle their lull of activity but also to maintain vigilance during highly repetitive monitoring tasks.

Finally, in Chapter 6, I will re-examine the notion of attention and discuss theories of joint attention and joint intentionality. I argue that the distinct human capacity to join in others' attention, as shown by Michael Tomasello's extensive research, is a useful framework to reflect on digital-mediated communication.

Chapter 1

Some Characteristics of Digital Communication

Communicating with an Intimate Network in 2009

By 2009, in developed countries, by and large, most people had access to multiple mediated channels of communication every day: voice calls through a fixed or mobile phone or through some VOIP channel; textual communication via e-mail; texting, messaging, and social networking platforms; and images and videos via social media in various forms. A good proportion of adults in Europe and the United States were using at least four different channels a day. Many regular Internet users had more than one e-mail account, and most social networking users had more than one personal page. In a number of European countries there were more mobile subscriptions than inhabitants.

All this is well known and extensively hyped. What is less known is that with all these channels, devices, and services, a user contacted on average the same five people 80 percent of the time.

The media rarely reported this concentration of exchanges on very few relationships, and very often users themselves were not aware of it. The concurrent emergence of social networking sites and the sudden generalisation of new forms of very ostensive public communication, obfuscated the more mundane transformation of private and intimate exchanges. And yet, even now, the vast majority of people have regular contact with only a tiny subset of the people they know, essentially with the inner core of significant relationships. What has significantly changed have been the social settings in which these exchanges are taking place.

Intimacy at Work: How Digital Media Bring Private Life to the Workplace, by Stefana Broadbent, 15–29.

In the following chapters, I will argue that the profound transformation of intimate communication has been no less subversive than social media and has in fact deeply challenged some well established norms and institutions.

Communicating with Loved Ones

Between 2004 and 2008 I ran an Observatory of Digital Life at Swisscom, the national Telecom operator in Switzerland, where, with a group of social scientists, we carried out extensive research on people's communication habits. Over a period of four years we asked more than six hundred people, from all age groups, life stages, professional, linguistic, and regional backgrounds, to keep a record of all their communications, excluding only professional exchanges and face-to-face conversations. Participants were asked to keep a diary for four days jotting down every mediated interaction—dialogues via SMS, e-mail, voice calls on landline and mobile phones, and IM sessions or calls from the PC.

It emerged that conversations and exchanges were concentrated among very few partners. In most cases the diversity of channels did not include a diversity of interlocutors on a daily or weekly basis. Multiple channels were being used with the same partners, depending on different situations and different content. Although on average the number of interlocutors mentioned in the diaries ranged between seven and fifteen, most of the exchanges were concentrated on five people. This was particularly true with voice calls from mobiles. Our qualitative data was then confirmed by a data analysis (Schnorf 2008) of the mobile-user database of 5 million subscribers, indicating that 80 percent of the calls from any one phone during a one-month period were made to four people. Written channels such as SMS and e-mail seemed to be slightly more diverse in terms of the number of communication partners.

We also found that channels are used redundantly: the more intense the relationship with someone, the more likely it was that a variety of channels to communicate with that person were being used, calling them on the mobile and landline and also sending SMS messages and e-mails. Parents, children, partners, close friends, and relatives were contacted frequently and with all the channels available.

By contrast, people who were at the periphery of the social environment—contacts seen less frequently or felt to be more distant—were contacted essentially on one channel only. The distant aunt would receive a Christmas card, the friend who lives far away an occasional text, an old colleague the occasional e-mail or message. Only exceptional occasions—such as a visit to the town where the contact lives, an important event, or a request for help—may lead to a change in the communication channel and an increase in the intensity of exchanges.

In 2009 a group of social science researchers from the Facebook Data Team published a report that analyzed the number of communication events between the total database of Facebook subscribers. It emerged that, on average, users had 120 friends but that they actively communicate with less than 10 percent of them.

Facebook researchers defined friendship networks in four different ways:

- All friends: the set of all people they have verified as friends.
- Reciprocal communication: the number of people with whom a person has reciprocal communications or an active exchange of information between two parties.
- One-way communication: the total set of people with whom a person has communicated.
- Maintained relationships: the set of people for whom a user had clicked on a newsfeed story or visited their profile more than twice.

For each user they calculated the size of their reciprocal network, one-way network, and network of maintained relationships, and they then plotted this as a function of the overall number of friends a user has.

People who had 120 friends (which at the time corresponded to the great majority of users) were actively engaging with fewer than 10 people. When the figures were divided by gender, it emerged that the average male user was leaving comments on seven friends' photos, status updates or wall, and messages and chatted with only four friends. The average female user was commenting on ten friends' photos, status updates or wall, and messages and chatted with six friends. In other words, Facebook users comment on posts from only about 5 to 10 percent of their Facebook friends.

Users were passively engaging (simply visiting their page or looking at their links) with between two and two and a half times more people in their network than they actively communicate with. The average user was regularly "checking out" the pages and posts of twenty friends. This passive engagement is what leads social networking site users to believe they are communicating more and keeping in touch with more people.

The PEW Institute for Internet and American Life Project, which publishes the most extensive and systematic data on the adoption and usage of digital media in the United States, presented a study in 2009, Social Isolation and New Technology, on the impact of ICT on people's social ties. Challenging the assumption that online digital life isolates people from their social network and neighbourhood, they found that online activity correlated with a greater level of local engagement and greater variety of social ties. People who regularly spent time online had a slightly bigger number of core ties and a greater likelihood to have nonfamily members in the core group of contacts. However, we should note that the difference was not great: on average the core network of social relations comprises three people—for active Internet users the average was 3.8 people.

This limited number of regular contacts was also found in countries outside Europe or the United States. Japan's most popular social networking site at the time was MIXI (Takahashi 2010), MIXI users had on average twenty people in their circle of MIXI connections, but of those users around 50 percent had four or fewer people in their "my-miku," and only 4.8 percent had over forty-one people. Again we found a very small, intimate set of friends being connected via these services.

Data from studies on women's use of mobile phones in China (Wallis 2008) and in India (Lee 2009; Pryanka 2010) also suggested that communication was focused on a few connections. The migrant women in China interviewed by Wallis had few contacts in their phonebooks. Overall Pryanka's study on gender and mobile usage in India indicates women's very restricted use of the phone in many rural households.

In 2010, wanting to study how communication patterns were changing thanks to services like Skype, my students in London and Paris interviewed people who have relatives and family abroad. The vagaries of recruiting people gave us a mix of households, some with very significant and recent ties with people in foreign countries and some who simply came from families who had migrated to Europe many years ago. We also interviewed some people living in hostels in London who were planning to spend just a few months away from home. We were completely blown away by how much communication is going on between adults and, specifically, their mothers.

The ages of the people we interviewed ranged from twenty-five to fifty-five, and we talked to both men and women, some living alone and others with partners and children. Their occupations were also very varied: students, job seekers, nurses, translators, bank clerks, and shop assistants. Some who were living in France or in the UK were originally from Spain, Algeria, Lebanon, Colombia, or Romania. Regardless of geographical distance—some mothers lived in the same city—we found that the most frequent, regular, and lengthy calls these adults were making were, systematically, with their mothers.

The conversations with their mothers were all oral, rarely if ever by text or e-mail, and took place either on the landline or on Skype, occasionally on the cell phone. Everyone mentioned a regular pattern of call, either daily or weekly. Women seemed to be calling on a daily basis, and men weekly or biweekly. The mothers were sometimes retired but sometimes still active professionally. Fathers were mentioned much less frequently; conversations with them were often incidental to talking to the mothers: on Skype, for instance, they might say hello when passing in front of the webcam. Interestingly, many of the participants mentioned that they spoke with their mothers more since they had moved far away than when they lived together under the same roof. They even considered that in many ways their relationship had greatly improved now that it was distant and mediated by technology.

But parents and children were not the only relations who saw their communication increase since the adoption of new communication channels (Brugière and Rivière 2010). In the United States couples talked, texted, or e-mailed each other many times a day, regardless of their age. In a report by the Pew Institute for Internet and American Life (Wellman et al. 2008), it emerged that 47 percent of married couples in America communicated on the cell phone at least once a day.

Jonathan Donner, a researcher who has been studying the adoption of mobile phones in Western Africa, describes the use of deliberate beeping or missed calls in Rwanda. These missed calls, in Donner's analysis, have different meanings: a request to call back, prearranged codes, or, very often, ways to say, "I'm thinking of you."

Patrick and his wife beep each other various times a day: "When my wife sees a morning beep, she knows I am just saying 'hi,' but if my wife beeps me twice in the late evening, I know that she is done with her work and then I always go to pick her up. I always call back if she beeps more than once" (Donner 2007, 13).

In our own research in France and Switzerland we also found that couples stayed in contact throughout the day.

Mireille and François

Mireille and François have recently had a baby. Mireille is on maternity leave and stays at home most of the day except when she goes out with the baby for a walk. In the bedroom of their small flat they have a computer (with a big monitor), which is always switched on and is their primary means of keeping in touch during the day. François's office is outside Paris. He has a long commute and also works long hours, often finishing late in the evening. The couple exchange a few e-mails during the day just to say hello and talk about what the baby is doing. When the evening comes and Francois is alone in the office, he switches on Skype and a webcam and so does Mireille at home. While he continues to work, they can have a chat, and François can see what is going on at home. Often in the evening, while the baby is sleeping, Mireille sits at the computer watching videos or e-mailing her friends and chatting away with François at the same time.

Chandy

Chandy is from Sri Lanka and had been living in Switzerland for thirteen years, for the last twelve working as a waiter in a luxury hotel. He has a wife and two children. Chandy at the time was a head waiter and highly appreciated in his job. He had recently been offered the chance of a second job in a nearby town, in an elegant café that belongs to the same chain of luxury hotels. He accepted in order to earn a bit more, even though it meant travelling thirty minutes each way by train during his "break" time from the hotel or working an extra shift at the end of the day and spending less time at home.

Chandy married his wife in Sri Lanka only a few years before, and she came back with him to Switzerland. She did not speak any German, spent most of her day at home with the children, and relied as far as possible on Tamil shops that she knew and a few Tamil friends. However, for most practical matters, such as arranging appointments with doctors or answering letters and paying bills, she had to rely on her husband. If anyone phones the house during the day, she would tell them that her husband would call them back later. Chandy phoned home regularly during his breaks to see if everything was fine and if there were any calls he should make or matters to attend to.

Because his wife really did rely on him for most things, Chandy's mobile phone was absolutely essential for him. In general he preferred to call her, and she knew not to call him while he was working unless it was truly urgent.

Once routines of mediated communication are established—such as calling each other during the day for a quick hello, messaging, e-mailing, or sending a picture or video to keep abreast of daily events—people tend to rely on these moments of contact and proximity. When, for different reasons such as financial or professional, the routines are interrupted and cannot be substituted with other means of contact, people can become very distressed.

Louis

Louis is a second-generation immigrant from South America who grew up in France and, although he still didn't have citizenship, felt very French. His parents worked very hard when he was a child and managed to put some money aside for their retirement and his future. They had genuinely suffered from being employed by people who did not consider them in the same way they considered French-born employees. That meant that having your own small business was the road to independence and safety, so some years ago they bought a small café and, after running it for a while, left it to Louis.

The café is cramped, and although in a very nice neighbourhood, it is on a small side street with little to no traffic. It is essentially a place for regulars—local residents who want to get out of the house. In South America going out for a drink or coffee with friends is a well-established respite from family life and home. In the small provincial town of G., where Louis and his parents live, if the habit ever it existed, it disappeared years ago. So business is very bad. Although Louis tried a variety of formulas to attract more customers—small hot snacks at lunch, happy hour drinks—nothing seems work, and he would have sold it ages ago if he didn't feel an obligation towards his parents.

At the time of the interview the situation was a financial catastrophe. His wife, Isabelle, had to stop working in the café and take on cleaning jobs to bring money to the household. Louis and his wife were always very close, either working side by side or talking together on their mobiles whenever they had a moment. By 2008 they couldn't afford two cell phones, so he gave up his and only had the café landline. His wife kept hers as a measure of security, but she could not call Louis during the day: one of her employers made a scene because she found Isabelle talking on her mobile phone during working hours, so now she calls her husband only when she is on her way home or between her jobs. To say that Louis and his wife were depressed is not an exaggeration; he felt terribly lonely and stuck in his failing café, and she felt alone in her jobs that she hated.

Louis and his wife had a very strong relationship that their financial situation strained. The unhappiness of both having jobs they disliked was made worse because they could not alleviate their loneliness with a short call or contact during the day.

Continuous Communication

Continuous contact with loved ones mediated by different digital channels rapidly emerged as fairly universal behaviour, embraced by people of different generations and cultures. It was also the first massive digital transformation we witnessed in mediated communication, largely preceding the social media phenomenon. Teenagers were the first to recognise the possibility offered by mediated channels for maintaining a sense of permanent awareness of their friends; they were rapidly copied by their elders.

Mimi Ito, whose research on teenagers in Japan was among the earliest studies of the phenomenon of mobile culture, described very vividly the feeling of being "within earshot." Writing about SMS, she says,

> These messages are predicated on the sense of ambient accessibility, a shared virtual space that is generally available between a few friends or with a loved one. They do not require a deliberate opening of a channel of communication but are based on the expectation that one is in "earshot" . . . people experience a sense of persistent social space constituted through the periodic exchange of text messages. These messages also define a space of peripheral background awareness that is midway between direct interaction and non-interaction. (Ito and Okabe 2005,132)

Her description of texting can equally apply to all the other channels of mediated communication, and what we have witnessed was, in fact, the use of multiple channels to ensure that this sense of being in a "persistent social space" is maintained throughout the day, across multiple situations.

Most of the exchanges we captured through our diary studies can be characterised as short updates. All through the day people seemed to make short calls or, increasingly, short texts, messages, and pictures to coordinate arrangements or simply to say what they are doing or feeling. These status reports, endearments, or sharing of some little event provide the sense of permanent awareness described by Mimi Ito.

When we interviewed Ron in 2008 he was in his late thirties and had recently had a baby with his second wife. He also had a teenage son from his previous marriage. He was the sole employee of a car rental company and had quite a lot of empty moments in his day while waiting for customers. The owner was an elderly gentleman who trusted Ron completely: he knew that Ron loves technology and staying in contact but that he is also serious and reliable when it comes to clients and cars.

Ron talked to and sent texts to his wife many times a day:

> I really like mobile phones—I have two and so does my wife. I have a mobile line with one company and one with another, and my wife has the same subscriptions as me so the calls to each other are free. Plus we have a fixed phone at home.
>
> I call my wife to say hello. I call my wife to know if everything is okay. I call my wife to know if the children are okay. I call my wife for the doctor. . . . I call my wife all day long! I'm a guy who really likes to participate, and I love hearing that the baby just ate his first apple or the older one just said something funny. And I want to hear it straight away, not tonight. Plus, I think it can be lonely for my wife alone at home with the kids. You know, she came to this country for me, and I don't want her to feel alone, so we talk as often as we can during the day. I really don't think she feels lonely this way, and if she isn't talking to me, she calls her mother and sister. So there is always someone with her, in a sense, sharing what is happening to the kids and sharing her day. She is my second wife and younger than me, and this time I am not risking another separation—it was just too difficult.

Ron couldn't wait for the kids to be old enough to talk on the phone so that he could be able to talk to them during the day. He and his older son, Jonas, were in touch with instant messaging (IM). The two of them had had a couple of rocky years, especially after he remarried and it was really difficult to talk to each other. One day Jonas told his dad that he had started using IM with his friends, and so Ron tried to chat with him on MSN. It turned out that this worked much better than talking on the phone, and since then that seems to be their main channel of communication.

> Now with IM I see if he is online, so I just say "hi" and see if he answers. I think it is because he can chat to me while doing other stuff he doesn't feel it is an obligation, a formal call from Dad. It's just in the flow of his activities. In a sense he is off-guard and so can tell me what is going on that day or that moment. And that is just what a parent wants, isn't it? Just knowing what your kid is doing and thinking and feeling that moment. You don't want a report; you just want to share a moment, be together a while. That's the hardest thing when you divorce, I think—the fact of not sharing as many moments. And somehow, it may sound absurd, but with IM, it's coming back. I really can be there with him for a while.
>
> So when I come into work, I put down my two mobiles. I open my MSN account, private e-mail, on top of the company e-mail and, obviously, company phone. I don't IM with my wife. I e-mail my ex because it is also the only way we have found to communicate in a civil form. Plus, if we do argue, no one hears us . . . haha.

Ron was unusual in 2008 in his desire to stay connected quite so often, but it became in the norm after people adopted smartphones, which enable a greater variety of channels than mobile phones. Reported data on usage now show averages of forty text messages and forty Whatsapp messages a day for adults in the UK (OFCOM 2014). Even in 2009, however, the majority of people we interviewed told us that they had daily or more frequent exchanges with some of their closest ties to share the little things that make up a day: encounters, small events, feelings, thoughts. The current growth of photo and video sharing sites, from Instagram to Facebook pictures, from Vines to Snapchat, are all part of the same phenomenon of updating the social surrounding of current states of mind. We will discuss in the next chapter the difference between these channels and how practices change according to the possibilities each system offers.

Personal Social Networks in 2015

Between the research mentioned above and the writing of this book, a few years have passed in which a plethora of new channels have emerged and been adopted and the impact of digital communication has been so profound in people's daily life that it is nearly impossible to think of social relations without considering digital media (Baym 2010; Boyd 2014; Turkle 2011). In a reversal of causality, some relationships are defined as central because the communication is frequent and intense. We could say that digital communication corresponds to what Mauss (1954) called a "total social phenomenon," a reality that permeates and defines nearly all realms of social existence. The advent of smartphones and constant access to social networking channels and messaging services such as Facebook, Whatsapp, WeChat, Weibo, or Instagram has changed and increased the frequency and intensity of exchanges. Statistics for messaging in particular are staggering in the number of monthly messages and the speed of growth. Whereas it was possible for us in 2008 to ask participants to keep communication diaries, it has now become impossible to expect people to write down when they are online, as it is a continuous state.

The ubiquity and continuity, however, do not necessarily mean that the network of regular contacts has been hugely extended. While the number of average contacts in Facebook doubled since 2009 (Bernstein and al. 2013 showed that half of the users of Facebook had more than two hundred friends) users still make a distinction between close online friends and acquaintances. Pew research from 2014 (Dugan and Lenhart 2015) showed that Facebook users in the United States only considered fifty of their contacts "real" friends even though their extensive online network included family members, old schoolmates, current friends and colleagues. Pew (Lenhart 2015) also reported that 70 percent of teenagers in the United States have at least some overlap of their friends on different social media. These findings suggest that we may now be witnessing two parallel phenomena: the intensification and diversification of communication within close bounded relationships, and the elaboration of new patterns of exchange with an extensive network of weak ties. The two processes may be happening concurrently but do not necessarily involve the same social mechanisms. In this book I am particularly interested in the former for the impact it has had in the redefinition of the boundaries between the private and professional domains.

Thus, it is useful to examine the composition of personal close social networks to understand how the concentration of communication relates to one's extension of the social sphere. When representing a person's social environment or ego-networks, it is common to draw it as a series of concentric circles. The circles closest to the centre contain the contacts the person feels closest to; those further out contain contacts that are less emotionally significant or less frequent. Not surprisingly, the communication partners on whom there is the strongest concentration represent the survey participants' closest ties: they are part of the inner circles of respondents' personal networks.

When asked to draw a map of their personal social network, respondents of all ages tend to write the names of the twenty people they feel closest to, and these for the most part are also the people with whom they are in contact frequently. Similar figures are mentioned in other studies undertaken in Western countries (Fischer 1992; Wellman 1979; Wellman et al. 2001; Wellman and Haythornthwaite 2002). Spencer and Pahl (2006), in their book on friendship, showed that the inner circle of most intimate relationships can include a mixture, in different proportions, of family members and friends. They can also include people who live far away, friends who have not been seen for years, or even important institutional figures such as carers or doctors.

Unsurprisingly, the relationship to the most intimate contacts also varies in nature: some friends or kin can be soul mates to whom everything is disclosed; others can be the sort of friends with whom having a good time is the basis of the relationship. Some people in the close circle can be those on whom you can count on for any event in life; others are just important reference points. Some contacts, as reported by Xinuyan Wang (2014), are recent friends who become significant because of migration or mobility. These relationships vary culturally too. In some cultures certain family or kin ties are by definition included in the most personal sphere of affects; in others more importance is given to "elective" relationships built over time.

Wang (2016) at UCL is carrying out a long ethnography in an industrial town in southern China, studying how rural migrants are using social media. Her research is part of a vast cross-cultural ERC project directed by Daniel Miller on the Global Social Media Impact Study (UCL). The project has field sites in China, India, Brazil, Turkey, Italy, and Chile.

Wang's informants are mostly young women working and living in a factory after their migration a few years ago from various rural areas of China. They all own smartphones and use them intensely to communicate, browse the web, take pictures, and play games. Wang's initial reports indicate an important difference from other areas, results from my own research and the studies on the communication patterns of migrants or transnational families. The young women Wang encountered seem to have fewer contacts with the families left behind than other migrant groups. Whereas the Colombian migrant women studied by Ramirez Angel (forthcoming), have Skype calls weekly or daily with children or parents left in Colombia, contacts between these rural factory workers and their family and friends who have stayed in their villages of origin are sporadic. The level of access to communication technology is similar or even, in many cases, better for the Chinese families, so the difference in frequency cannot be explained by a difference in digital access. What the young Chinese rural migrants explain is that they prefer to establish stronger links with their acquaintances or relatives in their same city or same professional environment. The families in the rural areas quickly become too different from themselves to allow for a significant mediated relationship to continue. The new acquaintances are more important, as they constitute their new proximal environment. This new social sphere is perceived as being more important for providing support in their new life and condition. Not only can they provide practical and logistic help, but they can far better understand their emotional states and challenges.

In the case of the South American migrants interviewed by Ramirez Angel, the link with the family is very close because often grandparents or aunts become the foster parents of the children left behind. Daily Skype calls, remittances, and the sending of gifts are all ways to parent from a distance (Madianou and Miller 2012) and contribute in maintaining a sense of control over the education and affection of the distant children.

Explaining the Emotional Intensity of Continuous Communication

The analysis of the personal ego-centred social networks of participants in our studies provide a grounding from which to understand the nature of the relationships being entertained thorough digital channels. A strong interdependence between the online and offline is confirmed in all the studies we have mentioned in the last few pages. The fact that people are essentially using digital channels with close contacts with whom they interact regularly through multiple media and that often these circles of ties are not very extensive is not in itself a huge surprise. In some cases the reverse is true: the intensity of the digital exchange generates the sense of proximity. There is considerable literature (Baym 2010; Boellstorff 2008) on the role of digital channels in creating strong emotional links that can precede or substitute face-to-face encounters. A recent study conducted by Maria Ramirez Angel and myself (2015) on chronically ill patients with limited mobility showed the strength of their emotional connection to other patients reinforced through digital media. Ethnographic studies of the Open Source Development Movement (Coleman 2012; Kelty 2008) also shows the strength of remote collaboration. What remains to be explained is the intensity, frequency, and continuity of the mediated exchanges: Why do people stay in touch so often? What happened in the last few years that led to an expectation of continuous contact? These questions are particularly crucial in order to understand how personal communication entered the workplace so rapidly and subverted some of the social norms regarding the separation of the private and professional spaces that had been entrenched for centuries. There are different layers of explanation of the phenomenon. In a following chapter I will discuss how the progressive privatisation, flexibilisation, and precariousness of work are pushing individuals to rely on their most intimate social sphere for support. This direction towards the personal, what Sennett calls the "tyranny of the intimate," provides a social and economic context to the phenomenon. On an individual level there are some conceptual tools of clinical psychology that provide some elements to help explain the strong association between safety and cell phones. Finally, an anthropological perspective allows us to situate these observations within a specific cultural framework.

The most suggestive ideas in the psychology of emotions, in my view, come from theories of attachment, which explain how the strong emotional ties people create in childhood will shape the way they handle significant affects and separation in adulthood. Classic theories of emotional development, such as Winnicott's theory of holding and good-enough parent (1965) or Bowlby's attachment theory (1973), detail the way children come to interiorise the secure environments offered by carers in infancy, thus enabling them to cope with separation and stressful situations as they grow up. As most of the contacts that happen through mediated digital channels are made during moments of distance and separation, it seems relevant to examine the emotional states involved in separation from the most significant relationships. In later chapters I shall invoke some historical and social transformations to examine why people are relying so heavily on the closest ties to obtain the emotional support they need, and why they cannot find the level of solidarity they need in weaker connections.

John Bowlby, who developed the theory of attachment in the late 1940s, started from his observations in children's hospital wards and in institutions in which children were separated from parents. He observed extreme cases of grief, withdrawal, and even impaired development in these isolated children. Invoking principles derived from ethology, he claimed that infants' instinctual behaviours of clinging, sucking, and smiling develop during their first year into an organised attachment to the mother or main carer. "Briefly put attachment behavior is conceived as any form of behavior that results in a person attaining or retaining proximity to some other differentiated and preferred individual" (Bowlby 1979, 129). Bowlby considered that attachment behaviour had a survival value (in order to preserve the baby from predators and other perils), which explained why it can be found in most mammal species. With development, the attachment behaviour decreases, and crying, calling, protests, and so on gradually become evident only in cases of extreme grief, fear, or illness. Bowlby and Ainsworth (1956) described various patterns of attachment that can emerge during childhood, ranging from the securely attached to anxious and avoidant. Securely attached children, with the sense that they have a secure base to return to, are able to explore the world around them. Toddlers can walk away from their mothers if they are sure she will still be there when they turn back to check. Children must develop a form of secure dependence on their parents before they can launch into unfamiliar situations in which they must cope by themselves.

Attachment, therefore, goes hand in hand with separation and anxiety. When separated from the mother, a child will go through phases of anxiety, desperation, and detachment. As children grow older they learn how to cope with separations and become more self-reliant but will still turn regularly to their carers for comfort. A secure base encourages the development of necessary skills for greater independence. Bowlby claimed that the responses of the mother or main carer to the infant—how available she is in meeting the needs of the child—and in particular their response to separation determines what type of attachment pattern the child and adult will develop. All through their life people will try to reduce anxiety by turning to the people they are attached to, and this, claimed Bowlby, is a healthy and adapted response. Bowlby observed that certain kinds of events, such as pain, hunger, alarming circumstances, criticism, or rejection by others, trigger anxiety in children and that children try to relieve their anxiety by seeking closeness and comfort from caregivers. A similar dynamic occurs in adults when events trigger their anxiety. Adults try to alleviate their anxiety by seeking physical and psychological closeness to their partners, as attachments in adulthood had been extended to partners and other people beyond the mother or carer.

Mikulincer and Shaver (2007) have developed a model for this dynamic in adulthood similar to what Tomkin (1962) described as scripts for positive affects. According to their model, when people experience anxiety, they try to reduce it by seeking closeness with their partners, who can provide support, including comfort, assistance, and information. When an event triggers anxiety, people try to reduce the anxiety by seeking physical or psychological closeness to their partner. The partner responds positively to the request for closeness, which reaffirms a sense of security and reduces anxiety so that people's normal life and activities can be resumed. If we substitute searches for closeness with communication activity, we see what many of the texts, calls, and other exchanges could be doing. The short calls or messages that we see dotting the working day and that are often directed to the most significant relations seem often to be there to reassure and have a supporting effect.

If we look again at what Ron said about the reasons for wanting to be in touch with his wife so frequently, it shows very clearly that he means to anticipate his wife's anxiety (of loneliness) and unwittingly expresses his own fears regarding separation: "I call my wife to say hello. I call my wife to know if everything is okay. I call my wife to know if the children are okay. I call my wife for the doctor . . . I call my wife all day long! . . . Plus, I think it can be lonely for my wife alone at home with the kids. You know, she came to this country for me, and I don't want her to feel alone, so we talk as often as we can during the day."

It is this impossibility of managing anxiety alone and coping with separation that, among other issues, Sherry Turkle challenges in her well-known book *Alone Together* (2011).

Identity and Communication

Another, more anthropological reading of the intense and frequent communication with a close sphere of contacts regards the fashioning of relational identities. Whereas the Chinese migrants interviewed by Wang are, as the women studied by Wallis (2013), in a trajectory of self-transformation, improvement, and modernisation, the women encountered by Ramirez Angel seem to want to preserve and control the social sphere they left behind. In both cases the way in which digital communication is being used provides a strong insight into the way different social identities are being negotiated. Wallis and Wang both have identified the role of the phone to support the process of refashioning the self from rural to urban, from low quality to modern and high quality. Wallis, because her studies were mostly carried out before the development and diffusion of smartphones, showed how the device itself was crucial as an act of consumption and ownership that moved the individual into modernity. Wang is seeing how access to multiple online channels and contents is refashioning the self. Sharing content with people in a similar situation of urban life contributes to creating a new social environment that is closer to the new subjectivity. Similarly, chronic patients talk of needing to find new sources of solidarity and identity in patient communities, as their condition alienates them from their existing social environment.

One of the claims I will make in following chapters is that work environments have been challenged by personal devices and private conversations because they introduce into the organisation a subjectivity that often contrasts with one's attributed professional roles. Kallinikos et al. 2013 explain that:

> an individual in a modern work organization is introduced qua a role and it is the role that becomes the object of managerial control and coordination rather than the person understood as a psychological entity with wider concerns and engagements (with family and community) that transcends work. (2013: 166)

The predominantly intimate role of smartphones makes another sphere of relations present in the working environment. The employee is there both as a junior staff member, for instance, and as a family member. The conversations that are held in the two roles are radically different, and the power relations in the private sphere have been subverted much more radically than they have in the professional space. One of the aspects that determines the difference in the relational spaces is the negotiation of attention.

Giving and asking for attention in communication is far from a neutral social behaviour and requires a sophisticated understanding of social norms and practices. Who we should give attention to first, for how long, how much attention we should request, and so on are highly complex social practices, and links between attention and power, attention and gender, and attention and education have been extensively studied. Charles Derber (2000), for instance, argues that status relations in general are fundamentally about the distribution of attention getting and attention giving across the social hierarchy. Those of lower status—including, in Derber's analysis, those who display more feminine behaviour—are expected to give attention to others. Those of higher status—including those who behave in the most masculine ways—are expected to demand and receive the attention of others. Thus, when either lower-status individuals give attention or upper-status individuals get attention, they confirm their status and their gender.

In workplaces the codification of attention is extremely explicit, and there are sophisticated artefacts such as online calendars, weekly meetings, appointment requests, and personal assistants to manage the request and attribution of attention between colleagues, subordinates and directors, clients and employees. The length of a meeting or of a call, the speed with which an e-mail is responded, the lag between requesting a meeting and obtaining it are all good indicators of the relative status of the interlocutors. The less attention someone receives, the lower their status or the greater the imbalance between them and their contacts. Digital media have done little to modify these social power interactions but have simply shifted the modalities in which they are enacted.

In contrast, communication between family members and friends tends to have resolved the process of managing attention allocation. Patterns of communication, with more stereotyped roles of attention giving or seeking are usually entrenched in the relationship and have given rise to regular habits. In some couples, for instance, one partner will always initiate the call, in some sibling relations one of the siblings will always expect to be called, in other cases any form of communication is welcome and acceptable. These habits simply show that the attention process has been resolved and crystallised in some habit. In the past these habits were ritualised in times and places in which calls would be performed; today the immediacy of multiple communication channels does not imply that any and all times are considered suitable. Transnational families often still talk of the Sunday call, the morning message, or the goodnight text. What is clear is that the patterns of these digital exchanges are intimately woven into the relationship as Madianou and Miller (2012) have discussed in regards to parenting from a distance in their study of Philipino mothers who have migrated away from their families.

The intense communication we witness with few close ties is, therefore, the result of a number of factors that point towards the use of digital channels as a significant means of enacting the relationship: being a good mother is Skyping every day, a caring spouse involves messaging regularly, a good friend means sending pictures or liking posts, tweeting daily is part of being an active professional. Intimacy is being performed digitally through access, availability, and constant visibility. In a certain sense the manifestation of the relation is displayed in the intensity and redundancy of the exchanges.

Chapter 2

How Digital Channels Are Supporting Intimacy

New communication behaviours are emerging from the constant recombination of multiple diverse channels and media. No single channel is responsible for all the transformations we are witnessing—it is the palette of written, visual, and oral channels that people use on a daily basis that are jointly contributing to establish new practices. The first social science books on digital communication focused on single channels or devices, such as the mobile phone (Castells 2006; Horst and Miller 2006; Ito et al. 2005; Ling 2004 and 2008), SMS (Kasesniemi 2003) or the classic study of the development of the landline in the United States (Fischer 1992). A few books on the combined use of multiple channels emerged later (Baym 2010; Madianou and Miller 2012; Ito 2010). Now, years of observations have shown us that most people use multiple channels every day with a certain element of redundancy. They typically engage with a range of social media, multiple messaging, e-mails, and voice services (Hampton et al. 2010). Users are highly sensitive to the subtle differences between channels, rapidly identifying the most appropriate social uses for them, and developing unique genres of communication. In *The Breakup 2.0,* Ileana Gershon (2012) poignantly described how students could be more offended by the choice of the channel used to communicate a relationship breakup than by the event itself. Being dumped by text or e-mail was perceived as a particularly cowardly behaviour by some students, just as being subjected to a constant stream of messages was considered intrusive by others. Boyd (2007a) famously reported the class distinctions between users of the social networking site MySpace and Facebook. Horst and Miller (2012)

Intimacy at Work: How Digital Media Bring Private Life to the Workplace, by Stefana Broadbent, 31–41.

describe how different modalities of digital communication come to be entwined with the expression of social roles, such as Skyping to be a good mother or texting to become a caring partner.

In 2008 (Broadbent and Bauwens 2008) we categorised the different channels based on the usage patterns we had observed over a period of four years. The categories sound somewhat old fashioned now that photography and video have taken such a major role in daily communication (Hand 2012), but the underlying criteria that led to the attribution of a channel to a certain function are still valid. When discussing specific instances of communication such as a call made the day before or an e-mail they had sent, our informants were able to clearly articulate why they had chosen a certain medium. They considered factors such as who they were communicating with, the type of content, where they thought their interlocutor was located, the tone they wanted to convey, when they expected a reply, and so forth. At the time, the realisation that channels were not substitutive and that cost was not the only selection criteria came as a surprise. For many years, before Internet services became available on smartphones, the assumption was that the emergence of a cheaper channel would somehow replace all the others. Skype in particular was perceived as a completely disruptive service that would kill traditional voice calls and messaging services. This did not happen, although it did radically modify international communication within families in particular. Although some channels do eventually disappear, the lag is long, and for extended periods we witness the coexistence of different generations of media. Currently SMS, for instance, still coexists with chat services, landlines with Skype, Facebook with Instagram, and letters with e-mail. OFCOM (2014) reports the decline of the number of SMS or the slow decline of the landline in favour of cell phones, but as of 2014, these channels are still in use, albeit included in a wider palette of services.

Similarly, the significant decline in the number of letters has not yet led to a complete substitution. What we do see is a reassignment of function that often reduces the spectrum of situations in which a certain medium is deployed, as other channels assume some of the communicative functions of the older channel. The specificity in the use of each medium, which partially explains their endurance, is a function of multiple factors. There are features that are integral to each type of channel: whether it is text based or voice based, synchronous or asynchronous; whether content is limited in length or duration; whether it provides information on presence or not; whether it enables multiple participants; and so forth. These features do not determine a usage pattern but do generate practices that seem to be quickly shared and adopted. They also come to be embedded in social norms regarding interaction, obligation, and access (Ryan 2006).

In our diary studies in Switzerland, for instance, we had found that the landline phone was being used for communication relevant to a collective, be it a household, organisation, or a business. Calls made from or to the landline were often relevant for everyone in the collective or at least for individuals in their role as household members or representatives of an organisation. Typically it was the preferred channel for keeping in contact with the social network of the family and for requests of services of a professional nature. In contrast, the mobile phone was dedicated to personal friends and contacts. Whereas calls to a landline were made to people in their roles as professionals or household members, calls on the mobiles were directed and made as individuals. It was perceived as a more personal channel, used for microcoordination with the closest sphere of family and friends. The sense of agency and independence that was expressed by teenagers or people in a socially controlled situation, when speaking of their newly acquired mobile phones, reflected this distinction. Mobiles in Europe, generally, were individually owned, seen as personal devices, and were dedicated to relationships held by the individual. As such, they supported private patterns of communication.

SMS was the channel most dedicated to intimate emotional exchanges with partners and friends. The content of SMS was more intimate than any other channel: the many short coordination messages were accompanied by numerous endearments and affectionate notes. Texting was highly personal, full of unspoken innuendoes, often funny and intimate. E-mail, by contrast, was used more for "administrative" purposes in support of online activities (e.g., travel and shopping preparation) or coordination with associations or administrations (e.g., receiving newsletters, organising and responding to bureaucratic requests). Friends and family were contacted by e-mail essentially when there were pictures or other digital content to be sent. The distinction between texting and e-mail reflected in some ways the one between the landline and the mobile voice calls. Texts were private, personal, and intimate, while e-mails were public and directed to organisations and collectives. Interestingly it should be remembered that in 2008 most e-mails were sent from PCs, which were physically situated in offices and homes, collective spaces by definition.

Instant messaging was emerging at the time as a really disruptive channel: it introduced the idea of being continuously in contact. Thanks to presence information and the fact that it could run permanently in the background, users had got used to the idea of having very long sessions and a sense of continuous companionship. Skype was also modifying the perception of time in phone calls, allowing for much lengthier and ambient calls. Used mainly by people with relatives or close friends abroad, it was extending the regular "Sunday call" to distant family, into a shared space where all sorts of activities could happen (cooking and Skyping, dining and Skyping).

At that time, social networking was in its infancy but was displaying some of the characteristics that were later confirmed by the massive adoption of social media. Social networks were already being used for visual content, to manifest a social identity, and to consolidate relations outside the immediate intimate and close circle.

The Dominance of Text

Since 2008 some of these channels have evolved more than others, and the devices themselves have extended the range of available channels. Although most commentators have focused on the exceptional adoption of social media, it is the phenomenal rise of text-based communication (according to the International Telecommunication Union, 6.1 trillion messages were sent in 2010 as compared to 1.8 trillion in 2007) that most modified individuals' communication patterns. Texting and messaging are, in fact, strongly related to the way personal communication became ubiquitous and continuous (de Gournay 2002, de Gournay and Smoreda 2001).

The unobtrusive and private nature of texting in particular allowed for "under the radar" exchanges to be carried out in public and professional spaces, radically expanding the settings in which personal communication could happen. Regardless of wider issues of surveillance, privacy, and security in relation to e-mail, Facebook, or texting, most people still consider text-based communication more private than voice simply because they cannot be overheard. As they were adopted, one after another, these channels have enabled a continuous flow of conversation in settings where private communication had been prohibited for centuries. In a following chapter we will discuss the impact that private communication is having on the workplace, but here I simply want to underline the fact that the combination of individual and personal devices and of text channels have allowed for a level of privacy that was not previously possible. In comparison, a landline or a letter are very public channels. Wallis (2013) and Horst and Miller (2006) report the privacy that new cell phone users in China or Jamaica felt after many years of having to use either a public pay phone or shared household landline. Calls and letters that reach a house or workplace could be picked up and answered by someone else, identify sender and intended receiver, and may be overheard or widely read. As discussed above, channels that reach a place rather than a person have evolved into "collective" channels dedicated to communication with the household or workplace.

The unobtrusive nature of text can also be seen in comparison to photos and videos. Unless performed in a private space, the overt nature of picture taking and video shooting make these channels fall into the public sphere. Taking pictures, be it of self, others, or situations, has a postural component that, for the moment, render the act itself deliberate and manifest. The visibility of the action, accompanied by a series of relatively normative bodily postures, especially in the case in which the video or picture are taken to be shared, make it a very ostensive way of communicating. The highly stereotyped selfies, pictures of food, pictures of pets and friends are not only an indication of the creation of norms of what can be shown but also of when, and where images can be taken. In a work setting, for instance, these types of communicational activities will be essentially performed in particular moments associated with well-codified rituals such as events or celebrations.

Decoding Text

From a cognitive perspective, it could be argued that the success of texting and messaging depends on the fact that they were used initially within bounded social networks. They became significant channels of communication because of the close nature of the relationships that were being entertained. The underdeterminacy and ambiguity of short texts requires a considerable amount of contextual information to be understood and, therefore, are most successful when people share a lot of prior knowledge. The 160-character limitation initially imposed by SMS meant that, as a channel, it was exceptionally informationally poor, requiring the interlocutors to use their contextual knowledge to make the correct inferences and understand the communicational intentions of their partner. This level of shared contextual knowledge is more easily found among people who share experiences and events and are, therefore, relatively close.

According to pragmatic theories of language (Grice 1989), communication is a process by which utterances and gestures produce relevant inferences that are supported by the context. In the inferential view, utterances are not signals but rather pieces of evidence about the speaker's meaning, and comprehension is achieved by inferring this meaning from evidence provided not only by the utterance but also by the context. In pragmatics "context" is not just the physical ambience of play; rather, it refers to all features of the perceptual and cognitive environment of the agent performing communicative acts, thereby allowing the hearer to infer the speaker's meanings (Sperber and Wilson 1995). Communicative processes are framed in environments that elicit a sense of relevance and the possibility of generating inferences that are meaningful for the participants, whatever the content might be.

Texting, therefore, was intrinsically suited for communication between close relations because it required a level of shared context to be understood and, in certain sense, excluded other more distant social spheres. In our studies we found that, even between close friends, when texting/messaging led to ambiguity (e.g., understanding whether the other person was joking or angry), they resorted to a "richer" channel such as a voice call to disambiguate the situation. It is interesting that Twitter has become a broadcast medium with the same limitation of character number. Many tweets, however, link to more extended pieces of text that expand the message. Also, as with much social media, tweets are often in reference to events that are socially and culturally shared on a wider scale through channels such as television or other Internet information sources. Tweets, therefore, can rely on the assumption that there is a concurrent context being constructed through other media, allowing for some level of mutual understanding.

Interestingly, even Twitter seems to accompany and extend face-to-face encounters. Takhteyev, Gruzd, and Wellman's (2012) study of the geographical distance between individuals in Twitter networks showed that a staggering 39 percent were in the same metropolitan area and that the best predictor of Twitter exchanges between two people was the presence of direct flights between the locations. Even on a channel such as Twitter—which is, in principle, predicated around centre of interests, independent of location and familiarity—the likelihood that face-to-face encounters either precede or follow Twitter exchanges is very high.

Asynchronicity, Attention, and Power

Another critical feature of text-based communication is its asynchronous nature. The social implications of this characteristic is that it allows people to manage the complex dynamics of relational attention. Giving and asking for attention in communication is far from a neutral social behaviour and requires a sophisticated understanding of social norms and practices. Who we should give attention to first, for how long, how much attention we should request, and so on, are highly complex social practices, and links between attention and power, attention and gender, attention and education have been extensively studied. As mentioned in the previous chapter, Charles Derber (2000) argues that the distribution of attention getting and attention giving across the social hierarchy are fundamental determinants of status. Thus, when either lower-status individuals give attention or upper-status individuals receive attention, they confirm their status and their gender. In mediated communication the negotiation of attention is expressed in the choice of channel.

Synchronicity/asynchronicity carry strong implications for the distribution of attention. Voice channels require both speakers to be communicating at the same time. Written channels are predominantly asynchronous, even when the time lag between a message and a reply is very short, as is the case with many current messaging systems. Synchronous mediated communication, as is found in voice channels, has a very strong prerequisite: both interlocutors must be available at the same time for the conversation and willing to dedicate the necessary amount of attention required. When people are face to face, it is easy for both interlocutors to see and understand whether the other person is available for a conversation. When people are distant, this readiness for conversation must be inferred or negotiated.

In a synchronous mediated communication the caller is aware of asking for attention and interrupting the other person's activities. Generally this type of request is not made lightly or unconsciously of the social implications. Respondents in our studies mentioned that they always hesitated to disturb someone at home or call someone on their mobile whom they did not know well. Most people will not choose a synchronous channel to communicate with a person in a very different hierarchical position, such as their boss or their teacher, but instead prefer a less intrusive mode like e-mail. Many sophisticated social techniques have been developed to negotiate this type of demand: sending a text or an e-mail first to ask whether a call can be made, fixing a set hour, or making appointments for phone conferences.

Interestingly, in our studies in Switzerland we found indications of gender differences in channel selection. Women and men explained their preference for text or voice with diametrically opposite arguments: women told us they liked being able to communicate without disturbing or asking for immediate attention; men expressed preference for channels affording an immediate response.

In other words, attitudes to synchronous or asynchronous channel selection were good indicators of the gendered nature of attention requests. In 2011 we interviewed two young women from Algeria studying in Paris who told us that they never called their brothers on their mobile phones but rather sent them a text and then waited for the brothers to call them. The brothers, who were also living in France, were an important point of reference and support to their sisters, who tried to talk to them on a regular basis. In sending a text rather than calling, the women were not trying to save money: they felt that it was up to their brothers to decide when they were available for a phone conversation. Their choice of channel corresponded very clearly to the relational asymmetry that they felt they needed to enact between them and their male siblings.

Different Levels of Obligation in Facebook and Cyworld

The lack of obligation to give and manifest attention is one of the elements that makes the broadcast functions of social media so entertaining: most people feel no obligation to comment or reply to a Twitter, Instagram, or Facebook post if it is not a direct message. Social media can be consumed, like television or any broadcast media, without the viewer having to do anything but enjoy what is offered. The lack of obligation comes from the semipublic nature of the communication; the high number of connections transforms the communication into a broadcast and, therefore, frees the recipient of feeling a duty to respond. As soon as the number of "friends" or recipients of the communication is reduced, the sense of obligation increases again—as could be seen in connection with the Korean social networking service, Cyworld.

Cyworld was the most popular Korean social networking service—in fact, the first social networking site to be widely adopted anywhere in the world. In Cyworld, buddies are named "ilchons," which is a much stronger metaphor than "friends." In Korean culture "chons" describe the closeness of kinship relations: one chon is the relation between parents and children, two chons is between grandparents and grandchildren, three chons the relationship with aunts and cousins.

In "real" life, where people cannot choose to form or terminate kinship ties, individuals are required to accede to the requests of others, regardless of how annoying the individual relatives may be or how burdensome their requests. In Cyworld, by contrast, users could start and stop Cy-ilchon relationships at will. As long as those relationships exist, however, Cy-ilchon buddies were socially obligated to participate in mutually reciprocal relationships and meet their partners' requests.

Some participants reported that they visited their ilchons' "minihompies" (homepages) on a regular basis and left comments to show how much they cared, a convention called "ilchon soonhwe" or "ilchon tour." Others sent each other "acorns" on special occasions, like exchanging gift cards offline. Cy-ilchons felt obliged to visit others' minihompies at least once a month and to pay return visits to ilchons who left comments on their minihompies. A special function of Cyworld, called "Surfing My Ilchons" or "Today's Ilchon," offers a convenient way for users to routinise random visits to some of their ilchons. As members of the Cyworld community, participants said they were aware of norms of the virtual world and tried to conform, believing that mutually reciprocal behaviours enhanced the sense of being connected to others for the purpose of growing "jeong" (a sense of mutual loyalty), bit by bit, among ilchons (Kim and Han 2007).

In the virtual world Cy-ilchons became as committed to each other as offline friends. A lack of return visits—in violation of the ilchon norm—indicated lack of politeness and neglect and often created relational problems. The norm of reciprocity made some users quit Cying for a while in order to be free of jeong burdens. This corresponds to what Hall and Baym (2012) called the phenomenon of entrapment, in which the sense of obligation to respond to text messages and calls from close friends can significantly reduce the sense of satisfaction with the relationship.

It is interesting to compare differences in the sense of obligation between Cyworld and Facebook. The rather loose metaphor of friendship that is present in Facebook, alongside the fact that the primary channel (the timeline or wall) is open to all friends to see, creates the fundamental lack of obligation. However, a more stringent selection process for buddies in Cyworld and relatively restricted access to the information that is shared (only cy-ilchons can see the homepage, which, therefore, starts to contain more private information) brings a sense of mutual duties.

In a study of 120 Cyworld users, Park, Heo, and Lee (2008) reported that 80 percent of them had fewer than 20 cy-ilchons, and only 5 percent had more than 40. Comparing these numbers to the average number of friends that Facebook users had by 2010—120, according to Facebook Data Team—we see a completely different scale of relationships. Of course, even among Facebook users there can be an implicit rule or sense of obligation to respond. In the teenage communities described by Danah Boyd (2007b), friendship is expressed by responding to postings and possibly solidarity demonstrated by responding publicly. In some circumstances failure to respond to particular posts can be read as a serious breach of trust, just as it is in Cyworld. Paradoxically, gaming environments can also produce significant obligations.

Farmville is a good example.[1] In this game, which became very popular in 2011, users build a virtual farm and rely on their neighbours (recruited among their Facebook friends) to help them grow their land. Being a Farmville neighbour required a certain level of commitment to regularly visit neighbouring farms and provide them with necessary crops and gifts. Groups of Farmville neighbours were closer in size to groups of Cyworld ilchons and had similar daily visiting rounds, in this case in order to ensure the survival of all the online farms.

People reportedly stopped playing this game when they felt the obligation was too onerous, just as we saw in some of the Cyworld users. More extreme and complex cases are those of participants in World of Warcraft guilds, where players must come together for prolonged periods of time to play jointly in order to progress in the game and combat other guilds. There, again, the obligation of presence is very high, and players are quickly excluded if they miss some sessions, thus putting significant pressure on guild members to participate, be focused, and demonstrate solidarity and commitment to their group, which again is relatively small and selected.

Generally speaking, however, in Europe the sense of obligation and duty to respond is still associated with one-on-one or small in-group channels such as e-mail, texting, or messaging. Hall and Baym's (2012) article on entrapment and expectations of communication between friends refers to texting and calling, two channels used for one-on-one communication. This is in contrast with what Naaman, Boase, and Lai (2010) called the low cost of engagement in weak-tie maintenance in social media.

The Shift to Small Group Communication

One of the characteristics of sociality in the European countries we discussed in the previous chapter is the presence of relatively small ego-centred networks. The people we studied seem to connect intensely and frequently to a small group, which they know well and with whom they share a combination of communication channels. Even children and teenagers, as shown by a recent Pan-European study (Livingstone. Mascheroni, and Murru 2014) have smaller networks than expected and concentrate their exchanges on people they know personally: 79 percent of friends on social media are personally known by the young people interviewed in the European survey.

The evolution of instant messaging services, such as WeChat in China (600 million users in 2015) or Whatsapp (800 million users in 2015, according to Statista) and Snapchat in Europe and the United States, go in the direction of supporting communication between these tightly knit circles. These services allow users to create a bounded group, which exchanges messages only within its boundaries. A message, photo, video, or link sent to the group is seen by all and only the members. The new generation of instant messaging services have emerged in the wake of the success and expansion of more open social media such as Facebook, Qzone, LinkedIn, or Twitter. A few years ago it became apparent that in between one-on-one and broadcasted media there was space and a demand for communicating effectively within more restricted groups (of typically between three and ten members).

Whatsapp presented itself as an easy way to build groups from existing connections with people of which one had a phone number. This characteristic distinguished it from the grouping functions offered by Facebook, for instance, because it tied itself to a mode of identification that usually implies personal knowledge of each other. The rapid uptake of Whatsapp and WeChat was also determined by the possibility of effectively sharing pictures, news, and videos within small groups. In this sense they widely extended the communication palette of previous texting and messaging services.

The success of these small group channels challenges some of the ideology of social networking sites, predicated on the "networked individualism" model described by Rainie and Wellman (2012). Networked individualism opposes a traditional model of sociability, that of being organised in tightly knit communities (of villages, families, or workplaces), to a loose set of connections centered on the self, which are opportunistically mobilised when needed. The members of the ego network are not necessarily linked to each other apart from the connection to the individual and, therefore, do not form a community. The absence of institutional belonging (to churches, workplaces, or clubs) as described by Putnam (2000) means that individuals form their own networks based on collections of life phases and experiences. This approach provides a powerful framework to understand how social networking sites allow individuals to collect, in a unique space, a large number of often unrelated contacts. It does not necessarily account for the relatively permanent groups of interrelated connections that coexist and cross-communicate within the group messaging systems.[2]

The greater privacy offered by group messaging compared to other forms of social media is usually invoked to explain its massive diffusion. The possibility of restricting the number of people who can view messages and content corresponds to a need for increased control on the circulation of information. With the proliferation of photography in daily communication and of images of the self, the exposure to the growing number of connections on some popular social media has raised concerns. Group messaging has therefore been presented as a more controlled environment in which individuals can be selective in who they share information with. This, as we saw in our discussion of Cyworld, comes at a cost, as it increases the obligation to reply and engage in the conversations.

Privacy and control are the most frequent explanatory frameworks used to analyze the phenomenon. To these arguments I would like, once again, to add a cognitive element. As said previously, communication relies heavily on context, and small, tightly knit groups can be expected to share contextual knowledge more than highly dispersed networks. In most cases it should be remembered that these channels augment or continue face-to-face conversations and are profoundly embedded in daily life and relationships. The constitution of bounded groups around interests, family ties, or friendship increases the level of shared contextual knowledge and, thus, augments the relevance of the information for each of its members. Moreover, the visibility of the exchanges among others provides additional significant information. Seeing how others respond or react to messages provides a different level of information that enables forms of acquisition of knowledge that have been studied in education and social psychology. The group provides "scaffolding" (as intended by Vygotskii 1978) to individuals in their understanding of events, memorisation, perception, and so forth. The greater emotional or epistemic proximity of the group members thus enables effective forms of transmissions of knowledge. It is not a surprise, therefore, that small groups are also using these channels to exchange topical information.

Chapter 3

Intimacy at Work

Workplace, school, battlefield, foreign country—until recently, in these settings people were removed for short or long periods of time from their closest ties, family, or friends. Clocking in meant leaving behind the family and its concerns; emigrating meant saying good-bye, maybe for years, to children or parents; going to war meant limiting contacts to occasional letters. Geographical distance was not the only reason for this isolation; social expectations were just as binding. While at work, at school, or on a mission, the individual was expected to concentrate uniquely on the professional environment, avoiding all distractions from a private life that would hinder or jeopardise the job at hand.

In Europe and the United States separation from family and close friends and commitment to the people surrounding you is still seen as an essential part of succeeding in any serious activity, a precondition for becoming a reliable worker, a good student, a trusted colleague.

Children are thus schooled beginning at an early age to separate themselves from parents and family and are taught how to focus their attention, avoid distractions, avoid chattering with friends, all while still creating strong and enduring ties with their classmates. The school, therefore, represents the first institution in which an oppositional model of private/professional is enacted. The main techniques to ensure a successful cleavage between the environments will subsequently be repeated in the workplace: physical separation between the home and the center of activity, rituals of entry in the environment, separation from other communities by markers (uniforms, badges, identity tags), control of entry and exit, schedules, rituals for group bonding (assemblies, house system, teams, competitions, hierarchies, punishment for lack of participation), and control of attention.

Intimacy at Work: How Digital Media Bring Private Life to the Workplace, by Stefana Broadbent, 43–59.

Against this backdrop the wide adoption of private mobile and smartphones has been an extraordinary social transformation. Research on new communication practices, mentioned in previous chapters, has shown repeatedly that there is a continuous flow of contacts during the day and that the vast majority of these exchanges, whatever the channel, are with the people who constitute the most intimate part of a person's social network. So we know that private communication from school or work is frequent and is happening between people with very intense emotional ties.

A Changing Work Tradition

For the past 150 years the relationship between work and home has been essentially oppositional. The most widespread work model required a rigid separation of private and work realms: workers were paid for dedicating their strength, skills, and attention for an agreed amount of time uniquely to the activities needed for the completion of their job.

The separation of workplace and home is relatively recent, arising with the Industrial Revolution of the nineteenth century. When the great mills and factories were built, workers had to carry out their tasks on specialised machines that they could not bring home. Machines transformed the production process; labourers had to work where the means of production were located. This progressively emptied the house of the work activities that had been carried out there, both on farms and in cities, where homes had often included small shops or workshops. Until the mid-nineteenth century in Europe, homes were as much a place of economic production as they were of shelter and reproduction. In cities with a medieval history, like Florence or Amsterdam or London, you will still see streets named after guilds and professions, survivors of the centuries when labour and daily life mingled completely.

From the Private World to the Workplace

This separation of realms has been embedded in the urban plans of modern cities, with residential areas isolated from manufacturing, administrative, and commercial areas. Most large cities have dedicated the city centre to offices and shops and grown a sprawling suburbia dominated by family dwellings. The birth of suburbia, which separated residential dwellings from the public spaces of commerce, production, and bureaucracy, also meant the advent of commuting. Although flexible hours and self-employment are on the rise, an integral part of the working day of most employees is the daily journey from their house to their workplace.

This journey involves physical relocation, but it is also a psychological transformation from the home persona to the work persona. In her book *Home and Work*, Christena Nippert-Eng (1996) described how people go through a set of rituals to move from the home mentality to the work mentality. The separation between home and workplace is not simply spatial: some people she interviewed were working at home. But the two environments correspond to two mental states and identities, and everyone has elaborate techniques and habits to shed the home "persona" in the morning and get into the work one and then leave the work one in the evening to resume the private one at home. These practices can be as simple as putting on specific clothes for each environment, changing shoes, drinking coffee when arriving at work, or having a beer at the pub at the end of the day.

It is clear from Nippert-Eng's observations that adopting the work mentality is far more difficult than shedding it in the evening. The fact that getting work-focused requires such an elaborate set of rituals indicates that putting "other" thoughts in the back of one's mind is emotionally costly. However, this huge effort of revving up attention and leaving behind personal issues is seen as a normal part of a professional stance. Three phenomena seem to be happening in the morning: people are putting on a face or persona compatible with their professional role, building up their concentration for the execution of their job, and, to this end, actively removing personal issues from their attention. This transformation is accomplished by small chores like tidying up before leaving the house or by separation rituals with family members.

In our own observations we saw that people listen to different types of music and radio channels on the way to work than on the way back, read different types of information and drink different drinks. In the morning, on the way to work, people would listen to the news and to spoken programs on the radio. In the evening they were more likely to listen to their own music on personal devices. Everything on the way to work is oriented towards building up focus, attention, and concentration; on the way back it is all about winding down.[1]

Another striking observation we made was the systematic call or message at the end of the working day (or more recently the process of going through a social media feed). In 2008 telecom operators recognised a peak of calls in the early evening (usually 5 p.m. to 7 p.m.) at a time corresponding to the moment most people leave their workplace. Our respondents' diaries showed that calls and messages, all around the same time, were undertaken on a routine basis. These were not simply calls to ask whether something was missing or at home or something needed to be done. They were daily calls that marked the transition from the professional world to the private world. These calls help people shed the professional stance and regain their private persona. In a sense the calls redefine the person who can quickly re-enter the family and the intimate role that was left at the door of the office in the morning.

Data on Communication at Work

There is little data on precisely how much private communication is taking place during working hours. Most existing studies have an alarmist undertone: they measure working time wasted on nonprofessional online activities. Few studies examine precisely the number of communication exchanges and personal communication in particular.

Once again the Pew Internet and American Life Project offers some of the most detailed data for the United States. In a 2008 survey, by Madden and Jones, on Internet and cell phone usage at work, the data showed that 54 percent of employees with a personal e-mail account checked their inboxes from work at least occasionally. Those who used their mobiles for e-mail, 50 percent said that most messages they sent and received were personal, and 25 percent said that they were split equally between personal and professional.

In 2010 Madden and Rainie (also for the Pew Research Center) found that American adults were sending an average of forty text messages a day (with a median of ten), of which 84 percent were just to say hello. The majority of people surveyed said that all their text messages were private, and only 2 percent said they were linked to their work activity. This does not mean that working adults were sending ten text messages a day from their workplace—several of these were sent from home or while travelling—but certainly a proportion was sent and received at work. On the voice communication side, Pew found that over half the people interviewed never used their mobile phone for work-related calls. More recently Statista (2014) reports an average of more than forty Whatsapp messages sent per day; however, there is no data about the proportion of these sent from the workplace.

A Global Pattern

Similar data emerged in Australia. A study by the Australian Research Council (Wajcman et al. 2008), logged the use of mobile phones by more than two thousand people. Their results revealed that only 12 percent of the 13,978 calls made were work related. The mobile phone was used overwhelmingly for contacting family (49 percent) and friends (26 percent). Of the 49 percent of calls to family members, for both men and women, the highest proportion were calls to the spouse (18 percent). Women were disproportionately likely to phone their children (13 percent), parents (11 percent), and extended family (12 percent). Employed men devoted 25 percent of their calls to work-related purposes, whereas for employed women the percentage is 14 percent.

Drawing again on the Australian phone log data, family (47 percent) and friends (43 percent) were by far the most common recipients of text messages, from both males and females. Within families texting between spouses (19 percent) constituted the highest volume of text messages, and those who are employed are more likely to text their spouses.

Jonathan Donner (2006) who studied communication patterns in Western Africa, reports on a study in Rwanda with 277 micro-entrepreneurs, in which he found that only a third of their calls and text messages were business related; all the other calls and texts were social and personal.

Our research in Switzerland for Swisscom between 2004 and 2008 revealed that very few employees were given a professional phone. Small companies in general did not want to have to monitor employees' calls from business phones to ascertain what was private and what was professional. Many found a simple solution: they gave their employees a small monthly sum of money to cover the few short work-related calls they made with their own private phone. By and large only those employees in commercial sales positions had a phone subscription routinely paid by the company.

When we asked people to complete a four-day diary of their communication exchanges, we found that private communication from the workplace amounted to around 20 percent of all exchanges over the four-day period.

We looked at the time of the communications. Texting appeared to be spread throughout the day, whereas voice calls peak at lunchtime and in the late afternoon, periods corresponding to the lunch break and the end of the working day.

Analysis of who was being contacted during working hours confirmed that it was the closest ties (partners, children, very close friends) who were most likely to be contacted between 11 am. and 6 p.m., with peaks at 3 p.m. and 5 p.m. Weaker ties (friends and relatives) were contacted mostly in the evening from home.

The picture that emerged from all these different studies, in highly diverse regions of the world, was roughly consistent. All the research confirmed that private exchanges from work were more or less systematic and that written channels, when we exclude professional e-mail, are more likely than vocal channels to be dedicated solely to private exchanges. Cell phones were sometimes being used for both work and private calls, but the different texting channels are definitely for private exchanges. Private voice calls were generally undertaken during longer breaks, such as lunch or at the end of the working day, whereas texting, messaging or e-mailing can be done discreetly at any time of the day.

The other very important finding was that, when at work, people did not contact anyone in their social environment: essentially they actively contacted only those who are part of their intimate circle. They tended to wait for the evening, when they were at home or while commuting, to communicate with weaker ties and acquaintances. This is compatible with what we know about the different forms of communication people engage with according to how distant they are: the further a contact is geographically and emotionally, the longer the exchange, because contacts are less frequent and more mutual ground has to be constructed. Closer relationships need less context building and, therefore, exchanges can be shorter and with poorer channels such as SMS or messaging.

Emotions at Work

Underlying all these research findings are the emotions (Oatley et al. 1996) that are being expressed and exchanged from the workplace. There is not a single working parent that has not told us how important it is to be reachable by their children, the school, or child minders. Parents tell us of travelling back home again if they realise they have forgotten their cell phones. Short calls or messaging with their children at some moment during the day seem to relieve the anxiety and guilt that parents feel about leaving them in the care of others.

In his 2002 book *Paranoid Parenting*, British sociologist Frank Furedi discusses the ever-increasing fears parents have with respect to the world their children inhabit. Safety issues regulate much of children's life in school and outside it and are driving families to keep children indoors more so than in the past and to be less trusting of the community they live in. The timely arrival of mobile phones in this climate of anxiety (Furedi 2009) has made them one of the main tools by which to control these fears by allowing parents and children to stay connected throughout the working day. Any attempt to block the use of this device, therefore, goes against a set of fundamental and primeval instincts.

The Divorced Father

Mr S and his wife divorced a few years ago. His two teenage daughters still live with their mother. Mr S was obviously desperately missing daily life with his children. When he came to our interview, he was with his seventeen-year-old, who was on her way to a shop where she worked every weekend to earn some pocket money. As every weekend he had prepared her favourite sandwiches for her lunch break. He confessed that he called both girls twice a day: once in the early afternoon when they came home from school and again, when he could, to wish them goodnight in the evening. The young woman was very understanding with her father. Two calls a day were a lot, she admitted, but she was aware of her father's suffering and of the importance of these conversations for his well-being. The afternoon call was made from the father's office, and he had no qualms about calling his daughters during work hours. The routine he had established was of such emotional importance that, I believe, he would have left his job had he not been allowed to make these calls.

Many working parents have described their routine calls in the hours after school—to check whether their children are all right, to be sure they are doing their homework, to sort out fights between siblings. This practice runs across all professions and companies. Where access to a communication device is limited or regulated, parents will take breaks in order to make these contacts. As the age of the children changes, the channels of communication may change as well: parents of teenagers told us how they adopted messaging, for instance, to communicate with kids who would never respond to voice calls.

Mother and Adult Daughter

We interviewed a mother and a daughter already in her twenties. Both women worked in offices—the daughter as a web programmer, the mother as a secretary. The two women would often engage in messaging from their computers at work—short chats just to say how the day was going or recount some little event that was happening in their workplace. While the daughter's job implied a great level of freedom with respect to Internet access, the mother was more restricted. The daughter had helped her mother connect to some proxy service that overrode the office firewalls. The mother's colleagues were well aware of what was going on and were happy to take advantage of the young woman's expertise: the mother sometimes asked her tech-savvy daughter for IT support for colleagues as well as herself. Both women told us how these short daily chats were keeping them very close.

Adult Children and Elderly Parents

Mr D and Mrs T are brother and sister. Mrs T works for a public administration with extremely regulated access to the computer and to her mobile phone, whereas Mr D is self-employed in a small service company and can basically do as he pleases. Their mother is elderly and deaf and cannot have a phone conversation. They communicate with her mainly by fax, but in an emergency the mother can call and explain what she needs, although she cannot hear their response. They have established a rather complex means to check on her, by which the brother sends her a fax during the day, and she either writes one back or calls to give news. The brother then sends a text message to his sister to relay any news. This elaborate process has become a routine and seems to give peace of mind to all involved.

Couples

Couples are another category that keep in touch throughout the working day. People who work night shifts or early morning shifts will, nevertheless, find a moment to wish their partners good morning or goodnight. In general people with nonstandard working hours suffer from the fact they cannot be at home for ritual activities such as breakfast or preparing for sleep: it has always been the main complaint of shift workers, along with the disruption of their sleeping cycles. Short calls at the times that correspond to the rituals they are missing cannot fully compensate for their absence, but they do seem to make people feel that they are in some way participating.

The Night Shift

Mr F works a night shift in a packaging company. Six days a week his work starts at 9 p.m. and ends at 5 a.m. He is, therefore, never at home to prepare for bed and fall asleep next to his girlfriend, with whom he started living quite recently. The shop floor has very rigid regulations about mobile phones and has video cameras all over the working area. However, Mr F, along with his colleagues, has found a few blind spots where the cameras don't reach. When he has finished packing a batch of products, around eleven, he calls his girlfriend to say good-night. She has come to expect this call and cannot really go to sleep without it.[2]

Mr R is a baker with a shift that starts at four every morning and finishes around noon. He is never home for breakfast. So at 8 a.m. every day he wipes the flour off his hands and steps outside to call his wife as she sets off for work. He likes to wish her a good day and hear whether there is anything he should take care of on his way back home.

In both these cases the separation is not very long, there is no real urgency to communicate, and, in the case of the baker and his wife, who are in their late fifties, there is not even the intensity of the first months of a relationship. In both cases, however, the call from work is regular and ritualised. It compensates for the major sacrifice that their working hours impose, namely missing a transitional point of the day.

Even couples on more similar schedules may exchange some message during the day. Respondents' diaries show that these text messages or e-mails are often nothing more than a "How are you doing?" or "How did it go?" But even small coordination communications, such as asking the other not to forget something, fulfil the same purpose of providing a sense of presence in the other person's environment.

It is important to state that these moments of intimate contact are not always welcome and are not always comforting. On the contrary, they can be perceived as a form of intrusion and control. However, the emotional response to these contacts, be it positive or negative, is always intense.

The Unhappy Marriage

Mrs G. worked a couple of hours away from where she lived with her husband. He worked mostly from home, and she travelled to her workplace four times a week. Between her commuting and her working hours, it made for a long day. Every morning at 11:30 her husband would call her—even though he knew she worked in an open-plan office—to reassure himself that she had arrived safely and was all right. She was quite embarrassed by this ritual: as soon as the phone rang, she rolled her eyes and said to the colleagues around her, "It's my husband again. I'm so sorry." She mumbled curtly into her phone and put it down as quickly as possible, usually providing very little information about what she was doing. No one seemed to mind the call and most people understood that her husband probably felt quite lonely working from home all day. She, however, seemed to enjoy her job very much and didn't mind commuting. Her colleagues reckoned that her obvious displeasure and impatience did not bode well for her marriage. A couple of years later she divorced.

Communication with loved ones from work is stimulated by isolation and solitude. Jobs that require people to be alone for many hours obviously increase the likelihood of using personal devices to keep in contact with others. We have interviewed many transport industry workers, such as lorry, tram, and bus drivers (and we will discuss in a later chapter some accidents that transpired after the use of communication devices in trains and aviation) as well as night cleaners in offices. Our interviews confirm that anyone who spends several hours a day within a driving cabin or other semi-isolated space tends to use a personal communication device quite regularly.

Working in Isolation

Mr C drove a municipal van carrying equipment for roadworks. Often he had to wait hours for the equipment to be ready for collection, so he spent a considerable amount of time in his van, messaging and text messaging with friends. At the time, before smartphones, he had bought a special device for instant messaging because it was cheaper than calling and still allowed him to feel like he was having a real conversation.

Mrs B worked as a janitor in a very large university. Each janitor was responsible for cleaning a different wing of the building, so Mrs B was basically alone for the duration of her shift. Every evening she would call her partner, don her Bluetooth earphone, and have a long conversation while she worked. She said that being alone in the big empty building was both eerie and depressing and that chatting while working really helped her get through the time.

Mrs D was a bus driver in a small town. Although she considered her job quite sociable because she was always seeing people she knew as they boarded her bus, she wasn't really allowed to speak to them. When the bus reached the terminus, she used the few minutes there to send messages to her partner and to her colleagues, stating,

> When you are at the terminus, you're all alone. So I usually use this time to call back or write a message to people who have tried to reach me because I don't answer the phone while I'm driving. But when we're at the station, I usually chit-chat a few minutes with my colleagues. . . . But we also write texts to each other. . . . Before, we used to use the busses' radio, but they changed policy about its usage... it's true that we used the busses' radio channel a lot! Sometimes you would hear "Silence on the line!" and someone would answer "Wait, I still want to add something . . ." and it never ended! It was terrible! [She laughs.] But like in other places, things have changed. . . . But we still communicate a lot, using our mobile phones.

High-Stress Occupations

Jobs relating to health and community services also give rise to frequent private communication. Nurses, crisis workers, and others undertaking tasks involving intense teamwork and swift reaction to external events will use down times to take quick breaks to message or make a call, or they will schedule calls during appointed rest times.

A community worker involved in crisis management for drug abuse told us how everyone in the unit used their down moments when no new patients were arriving to engage in private communication.

A nurse who works in hospital surgical areas told us that during long operations nurses may have to wait for hours before they are needed again but still have to be at hand. No mobile devices are allowed in the operating areas, but there are some computers used for different administrative tasks during operations, and occasionally the nurses use them to send private messages while they wait.

Far from being signs of callousness or selfish disengagement, these calls or messages seem to help relieve the stress of some of the situations the workers have to handle.

Putting Emotions Back Together

The speed at which private mobile devices have been adopted and the systematic use of them at work are an unequivocal indicator of how tenuous and artificial the boundaries between private and professional were. Given the possibility, most people have immediately abandoned the belief that they could not include some form of contact with their private sphere during their work day.

The very strong association between safety and mobile phones was one of the main drivers of transitioning personal mobile devices in the workplace. Already in 2006 Castells discussed the relationship between safety and cell phone ownership, and Dutton and Nainoa (2002) showed that sales of mobile phones increased significantly in the United States in the aftermath of the 9/11 attacks.[3] In Europe safety has always been mentioned as a reason for having acquired a mobile in the first place and is systematically used to explain buying a mobile phone for one's children. There are a number of interpretations that can be provided to discuss the need for safety and proximity, from a psychological one that makes reference to theories of attachment and the strategies to overcome the anxiety of separation (Bowlby 1973), to sociological discussions of the individualisation of society, to social network analysis such as Rainie and Wellman's (2012) descriptions of small, interlinked networks. What is clear is that the popular discourse on control and safety facilitated the entry of such devices in the workplace in order to receive emergency calls. Compared to other far less successful initiatives designed to reduce the distance between workplace and home, (e.g., workplace nurseries, Marchbank 2004), mobile phones were allowed in offices at exceptional speed. However, it was their function as an instrument to receive emergency calls that was tolerated; their use to actively engage in personal exchanges was more controversial but eventually became the object of explicit or implicit new social norms of behaviour and practice.

In order to grasp how it was possible that these boundaries were redrawn so quickly, it is useful to refer to some sociological literature on the private-public issue. Sociologists such as Simmel (1971), Habermas (1962), and Norbert Elias (1939) and, more recently, Sennett (1977, 1998) and Putnam (2000), have all described the slow overinvestment in the private sphere and the progressive retreat from the public sphere.

The historian Philippe Ariès spells this out very clearly:

> When the city deteriorated and lost its vitality the role of the family over-expanded like a hypertrophied cell. In an attempt to fill the gap left by the decline of the city and the urban forms of social intercourse it had once provided, the omnipresent family took upon itself the task of trying to satisfy all the emotional and social needs of its members. Today it is clear that the family has failed in its attempt to accomplish that feat, either because the increased emphasis on privacy has stifled the need for social intercourse or because the family has been too completely alienated by public powers. People are demanding that the family do everything that the outside world in its indifference or hostility refuses to do. (1977, 2)

One of the most frequently debated phenomena in sociology has been the process of individualisation and disengagement from the public space that happened in Europe between the eighteenth and twentieth centuries (Aries and Duby 1985). Although there are opposing views regarding whether, for instance, individualism preceded the Industrial Revolution (MacFarlane 2002) or was caused by it, there is a certain agreement that families have grown to be the main source of solidarity and mutual emotional reliance, to the detriment of wider social arenas. Sennett, in *The Fall of the Public Man* (1977), talks of the tyranny of intimacy, on how we measure society in psychological terms. Like Ariès, Sennett describes how intimacy has become the most important aspect of human experience and how this has coincided with the demise of the public sphere (the city, the civil, or public society).

Paradoxically, in the last century, while we developed a culture that elevated the family and intimate life to the main social space of individuals—a space providing all the comfort, sustenance, and happiness that the external world cannot offer—people were made to work and learn in environments that cut them off from their families. From an emotional perspective, therefore, people were living in a society that overemphasised the role of intimate relationships and throws onto the family the ultimate responsibility of caring for and nurturing the individual while also spending a large portion of their day severed from those relationships. It is not at all surprising that the moment a channel of communication emerged that could be used to reconnect the two realms it was immediately adopted. By calling home during the day, people are simply connecting what has become their only secure bond.

The instability of work and the increased flexibility of jobs have only amplified that need. Nothing in the current economic environment suggests that belonging to an institution, whether a company, administration, or educational institute, is going to provide a secure and permanent identity and affiliation. Maintaining a permanent awareness of one's most intimate group is, therefore, perfectly consistent with the social priorities of our times. A recent study by Facebook's Data Team, Burke and colleagues (2013), has shown that, contrary to the long-held belief that weak ties are the most important source of information for finding a job (Granovetter 1973) it is the close ties that are the purveyors of relevant contacts. Our own recent research on the use of digital channels by precarious workers also supports the view that as people's work life becomes more fragmented and unstable, their social networks are reduced and consolidated around a small, close social sphere.

It is in my view impossible to understand the increasing reliance on close networks, family, intimate friends, and relations without considering the progressive fragmentation and instability of individuals' work life. In Europe the progressive erosion of social safety nets and welfare accompanied by a rise in unemployment are profoundly transforming individuals' relationship with the world of employment. We are witnessing a rise in self-employment in the UK, for instance, as a consequence of the reduction of employed jobs (ONS 2014), and there is also a reported increase of temporary and part-time jobs—all conditions that tend to be accompanied not only by a reduction of income but also of continuity in employment. Personal professional trajectories are more discontinuous, and transition phases between jobs are more frequent. In all these cases, be they personally experienced or simply feared, the intimate social network plays a major role in ensuring the continuity both in terms of identity and financial security.

Personal Communication and Precariousness

The recent analyses of the effects of the economic recession in Europe on inequality, growth, stagnation, and unemployment tend to examine ICTs for their role in the transformation and substitution of labour (Baumann 2006; Brynjolfsson and McAfee 2014; Kallinikos 2011; Scholz 2013). At the level of the individual, however, the dominant public discourse on the digital revolution systematically stresses the empowering dimension of the Internet to access networks of knowledge, information, and resources. Three years ago, with Maria Ramirez Angel at UCL, we started a research program on the digital practices of people hit by the economic recession in order to examine how online resources were being mobilised to support individuals and groups to adapt and potentially overcome the crisis. After a year of research we realised that we were, in fact, observing a far more long-term transformation because the emerging practices were reflecting new states of existential precariousness. This is a new condition in recent European history, characterised by an extreme professional instability, limited access to capital, and decreasing state support, leading to significant economic insecurity. Guy Standing (2011) describes a new class, the "precariat" as a class in the making: people who have fragmented work trajectories, social relations that are scattered over distant locations, and who lack prospects of stability. This insecurity is accompanied by a highly individualistic mode of relation and solidarity to small egocentric networks as opposed to more traditional professional or geographical communities.

All of the people we encountered during our ethnographic work fit into this description, and their use of digital media not only exemplifies this reality but also makes it viable. Digital communication channels are systematically being used as safety nets to compensate for the absence of more traditional institutional referents. Frequent and intense digital exchanges with a small intimate network of close relations ensure a minimum level of economic, emotional, and practical support. This process of enclosure is in complete contrast with the dominant discourse of empowerment through digital access.

Between 2011 and 2014 we interviewed young Europeans, often with university degrees, who are moving to London in great numbers from Spain, Italy, Greece, and Ireland in search of jobs and opportunities; South American adults who arrive in London in a second wave of migration after losing the stability they had acquired in Spain following the repossession of their homes or prolonged unemployment; and, finally, young homeless men from Eastern Europe or the UK sleeping in temporary shelters offered by a church. This apparently disparate collection has in common the fragility of their condition and their systematic use of digital media. This is a far from homogeneous group in terms of education, nationality, or social background, but the experience of prolonged uncertainty, which starts well before the arrival in London, determines a profound disruption in the construction of a life narrative accompanied by a systematic struggle to maintain a sense of agency.

The Promise of the Digital

The rhetoric surrounding the revolutionary nature of the Internet to empower underprivileged social groups by giving them direct access to resources that were until recently unreachable without institutional mediation has given rise to a wide range of initiatives. From MOOCs to online administrations, from projects to reduce digital exclusion to open government, there is an increasing range of platforms to support education, job searches, relocation, access to housing, benefits, and various types of community actions. The common underlying paradigm is one of access and participation, leading to self-fulfilment, autonomy, and economic expansion.

Horst and Miller, in their 2005 study of mobile phones in Jamaica, showed how these devices had become an integral part of economic survival for a large section of Jamaican society, where there is little or no accumulation of capital within households. Individuals would maintain through the cell phone an extensive network of ties that could be contacted for different levels of economic support when a need arose. This finding contrasted with the common discourse on the role mobile phones are playing in economic development in emerging economies. Rather than supporting new forms of entrepreneurship, cell phones were enabling people to cope with individual crisis situations through small transfers of capital.

In our study we saw similar forms of reliance on close networks that are accessed through different digital media, thus also challenging some of the received ideas about the power of the Internet to provide access to educational, institutional, and professional resources. The majority of our informants are digitally active and, in many cases, could be called "digital natives" but rarely if ever access local online resources for training, housing, and employment. Their exclusion from these online resources stem from a lack of language, social network, or cultural awareness and not from lack of access or skill. Their use of the Internet is, as van Deursen and van Dijk (2011) describe, centred on communication and entertainment and tends to reinforce rather than reduce the divide with more dominant and established users who are able to profit from the information and services made available online.

Conversely, digital media allow these users to maintain a constant connection with the intimate social sphere that, in this economic moment, is most likely to provide a minimum level of support. The homeless we have interviewed seem to be in such a vulnerable situation when they are missing or have severed ties of this nature. Digital media also contribute to maintain a sense of control and agency on certain aspects of life that could easily be swept away by mobility and instability. Access to content such as music, news, videos, information about less proximal relations, and consumption all contribute to maintaining a sense of identity and continuity. The greatest risk of "liquid life" (Bauman 2006) or the flexible regime (Sennett 1998) is the loss of a thread across experiences. Online access to people, cultural content, and memories support our informants in maintaining a sense of identity within the fragmentation of their trajectories.

Staying in Touch

In the groups we have studied, digital media is systematically used to maintain contact with a small circle of close ties. We have observed young people who live for months in hostels, lying on their beds Skyping with friends and family, or Colombian mothers using their mobiles to control the movements of their children. Through a variety of channels, be they Whatsapp, text, Skype calls, or international phone cards, all our informants have regular and frequent contact with families abroad, friends, and strong ties. There is ample literature on the use of social media in migration, parenting through Skype, and Facebook (Madianou and Miller 2012; Miller 2011, Wallis 2013) as a means of maintaining contact with families left behind. We have witnessed not only similar communication patterns but also the intense use of one-on-one channels to compensate for some of the vulnerabilities of the precarious situations in which the informants seem to live. Smartphones are, for instance, being used to check on children who are left home alone at night while parents work night shifts. Texting or messaging is used to summon a friend to a workplace for an immediate job vacancy, daily schedules are communicated to temporary workers through texts, and small services are received and offered via Whatsapp.

In case of extreme vulnerability or mobility the mobile/smartphone becomes the only identifying channel: the only point of contact for social relations, jobs, or link to public administrations. In absence of a home, stable address, or permanent physical space of any kind, the phone is the home, as many informants have told us. As such, it can be a source of discomfort as well as support, and we have witnessed many cases of avoiding calls or exchanges that are perceived as stressful or painful.

The precariousness of both employment and housing means also that the social relations established as a young adult, outside the realm of the family and school, are very temporary and transient. Social media are used to collect and store these contacts and relations as Facebook friends, e-mail addresses, or Instagram followers, but they do not form the core of the communicational space. In previous research (Broadbent 2008, 2012) we had shown that the majority of exchanges on mediated channels could be accounted by a small circle of close ties and that social media platforms were dedicated to exchanges that were more occasional and with more distant ties. This seems to be even more the case in a situation of instability, whereas the contacts accumulated on different social media platforms can only rarely be mobilised for support. However, the simple visual presence of one's extended social network, as is manifested in Facebook's or Instagram's list of friends and the potential access to this network, is perceived as providing a sense of continuity. Also having someone at least as an online contact reduces, according to our interviews, the sense of loss that accompanies the frequent separations.

The personal network seems to be the only social network recognised as providing a safety net in a condition of extreme uncertainty and is, therefore, constantly nourished and cherished and becomes the main source of information, material and practical support, financial aid, and emotional contact. In the case of the precariat, the effect of concentrating exchanges with close ties can have even greater detrimental effects by reducing access to sources of local networks and jobs.

Finding a Job

The increased individualisation of the people in precarious conditions means also that the more traditional trajectories of migration, which included family networks moving to the same locality and the role of religious institutions such as churches or community associations to support newcomers, are less prominent. In our research we have encountered many people who arrive in London with minimal preparation and network, relying on dubious relocation agencies or an accommodation booked for a few days. Regardless of the possibility of using numerous online sources to organise a landing period, most of our informants seemed to have scant knowledge of local employment regulations, the job market, the housing situation, health services, or even the topology of the city.

Looking for employment is the main reason to move for most of the people we interviewed, and London is seen as a city where jobs are plentiful and relatively easy to encounter. Employment laws in the UK allow for a degree of flexibility that is mostly impossible in other EU countries and supports a far greater market of temporary jobs. The process that is followed to find these jobs is extremely traditional and, in some sense, haphazard. There is a very limited use of online resources, both generic job hunting platforms and company-specific careers resources. Most jobs are found through personal contacts, acquaintances, people met in hostels or in church, or by leaving a written CV in a shop or café. The young Europeans who come, often after a university degree and a period of unemployment in their countries of origin, seem to focus their research on service positions—essentially shops, bars, restaurants—while the new Latin American community is more likely to be recruited in the cleaning business. Both groups have in common an insufficient mastery of English to allow them to look for employment in their area of education or experience. Leaving a CV in a café with a phone number may lead to a call if the business is understaffed. Cleaning jobs are instead organised by small businesses, often run by other Latin Americans, who leverage on lists of contacts that are called when needed.

Jobs themselves are often temporary and insecure, and this leads to the accumulation of positions. We have met many informants who have two or three jobs a day, adding up hours both to increase their income and fend off the risk of being without a job on any one day. The jobs are never in the domain of expertise or previous experience, as the insufficient mastery of English corners them into service positions that require minimum skills and reduced interaction with the public. For the young Europeans, this is seen as a necessary transitional phase, allowing them to improve their language competence. For people who are facing their second migration (from Colombia to Spain and then to the UK), it is a hurdle that is not always overcome.

It is in this area that the divide between the precariat and the more stable working population is most deeply felt when it comes to mobilising online resources. Knowledge about where to look, how to apply, what to put forward, and how to link to people are all crucial to profit from the digital platforms.

Alternative Economies

In Colombia in particular and Latin America more generally the concept of *rebusque* refers to the myriad of small side activities that can be done to supplement one's income. In Colombia this can range from selling phone cards to cooking and selling certain dishes. In London, within the Latin American community, there are many examples of *rebusque,* where individuals provide small services for compensation in cash. These exchanges of services run on texting (SMS or Whatsapp, primarily). There is no online presence or physical location, but contacts and advertisement are made through texting. Someone working in a market may text contacts to say he has received certain pieces of meat, a woman may text that she has made a certain type of sweets, a family may contact an acquaintance who can repair a computer or a fridge, and so forth. Similarly, the homeless we have encountered may share information via text about a new shelter or a place to wash clothes.

Shriram Venkatraman (2015), who is carrying out his research in Panchagrami, India, in the context of the Global Social Media Impact Study at UCL, reports of educated women using their smartphones and Whatsapp connections to offer different type of services after childbirth. Young IT professionals who leave employment after their first child start small home-based businesses relying on Whatsapp to advertise their services, such as preparing snacks for children.

The digital channels here merge with the informal extended networks of relations to support an alternative economy of service. This alternative economy provides numerous advantages to its participants, both financial and professional. When we consider that the majority of the people we met are not employed according to their experience and degrees, these side activities often constitute the only space in which previous skills are performed. Carpenters or builders whose official employment is as cleaners offer building services on the side on weekends, computer scientists offer repair or programming services, and so forth. Even the women interviewed by Venkatraman are finding new forms of professional identity and entrepreneurship in their proximal digital connections.

Precariousness and Digital Media

Dealing with fragmented professional trajectories and unforeseeable economic conditions is expected to increasingly become the norm for most people relying on work income rather than capital. Although professional and economic instability is neither new nor unique to Europe, what is different from the past is the social framework in which it is emerging. Different causes, be they globalisation, new forms of capitalism, and so forth, seem to be eroding some of the social institutions that provided a degree of resilience. Digital media, social media, and information services are all being mobilised to compensate for this uncertainty. They are doing so in an unexpected and somewhat dichotomous ways, on the one hand by providing some form of continuity in maintaining personal relations and cultural experiences, and on the other, by enabling access to short-term opportunities. The latter means that we are seeing emerging in Europe some of the patterns that we had associated with mobile phone usage in developing countries: the phone as a contact for occasional work or as means to access people who may help in times of acute crisis. Continuity, instead, is being offered by the myriad communication channels that allow permanent contact with a core set of relationships and by the possibility of accessing on demand the content that constitutes external memories of the different fragments of existence.

Chapter 4

Communication, Productivity, and Trust

The adoption of different communication channels to stay in touch from work has been both a gradual and spontaneous process. People simply started bringing their mobile phones to work and connecting to private webmail addresses and their social networking sites when they were able to. This social innovation came from the bottom up. It was not designed into the technology nor planned by the telecom companies and web services—it just seemed to happen when people realised the potential. The practice did, however, encounter some resistance, as companies and administrations started restricting access to the Internet and people's personal devices. It is important to analyse the historical background underlying the separation of individuals from their private sphere because many of the reactions, rules, and regulations that we will discuss in the following sections stem from this deeply rooted principle. The rules of separation are still present in schools, and some workplaces inherit nineteenth-century approaches to discipline and productivity. Zaretsky (1976) argues that the capitalist economy of the nineteenth century needed to isolate individuals from their families in order to break the family as a place of both reproduction and economic production. Thus, the isolated individual worker became a resource of a larger economic entity. The family as a haven, says Zaretsky, supported the capitalist model of labour by caring for the worker, providing the emotional and physical support that renewed him and every evening made him ready for the next day of work.

The rational organisation of labour introduced strong links between time and productivity. This, in turn, led to regulated breaks and working periods and put

Intimacy at Work: How Digital Media Bring Private Life to the Workplace, by Stefana Broadbent, 61–77.

the focus on control of workers' activities so that no productive time should be lost. Foucault's (1975) account of the evolution of the functional space and functional time in factories, schools, and military organisations describes brilliantly the onset of a philosophy of control of attention. Every minute, says Foucault, not only should not be lost but can be used to extract even more useful forces.

As management sciences have evolved over the years, we have witnessed a movement from focus only on the execution of the task to focus on the collaborative aspects of the job. The increasing role of teamwork, networks of knowledge, and co-operation does not alter the fundamental belief in the need to control attention. What it has done is to shift the focus from the machine or material to be transformed onto the social environment of the workplace in order to foster productivity.

Private Communication versus Productivity

The most frequent argument against private communication at work revolves around the relation between attention and productivity. In the absence of any regulation, it is felt, people would indulge in long periods of private communication and entertainment at the expense of productive time. However, analysis of the data on time spent at work in personal communication and the spread of it during the day makes it clear that this is not a fully rational argument: the issues at stake have far more to do with control, hierarchy, and the perception of work itself.

The productivity argument also looks highly socially selective, inasmuch as it seems to apply to lower-level jobs far more than to creative or managerial jobs. Companies that employ people with higher educational levels tend to be less restrictive in their rules regarding Internet or mobile phone access. More hierarchic organisations that have a tightly supervised production system based on tasks rather than autonomy tend to be more restraining. The unconstrained use of personal communication channels is, therefore, significantly dependent on the level of trust associated with a position. The real question lurking behind the rules is therefore the following : who has the right to manage their attention?

Reactions

Although not all institutions react the same way, initially there was a clear trend restricting the emerging communication practices. Schools are still generally limiting the use of cell phones, with measures ranging from confiscation to fines; many companies initially were blocking access to certain sections of the Web such as social networking sites and some still do; transport agencies are dismissing personnel found with mobile phones on the grounds of safety; shops, restaurants, and manufacturing plants still systematically prohibit the use of personal communication devices during working hours. The nature of the communication—and the importance of the relations that these exchanges are sustaining—however, means that a simple system of enforcement is bound to fail, and we have systematically observed employees finding ways to openly or covertly circumvent these restrictions.

Tom

Tom works full time for a company that makes coffee capsules. He recalls an episode from a few years back:

> At work there are no official rules concerning mobile phone usage anymore. Some time ago our boss decided to forbid the use of mobile phones during working hours. He put notices on the walls to inform the employees of this new restriction; we were expected to switch the phones off or leave them in our lockers. We all refused and told him that we were ready to do some form of industrial action if he didn't change his mind. You know, people have kids and want to be reachable. Now, I do get some calls and messages at work; I just keep them short and then call back during break time.

Kaspar

Kaspar was employed as a metal worker on building sites. On the sites the foreman decided whether having a mobile phone was or was not permitted. Whatever the rule on a site, Kaspar always carried his mobile with him, and although he was careful not to spend too long on calls at the expense of his work, he had no intention of giving up his phone. His daughter liked to call him sometimes just to say hello, and his wife called to ask him practical questions. Whatever a foreman may say, these calls were too important for him to relinquish:

> I bought a Bluetooth speaker so I could use my mobile hands-free. Before that, it was difficult to call while operating a machine, and I was more or less obliged to respect the rule of the boss and to call the least possible. The hands free kit allowed me to phone as often as I wanted and work at the same time.

Rose

Rose teaches secondary school children between the ages of eleven and thirteen. The teachers have devised a system to control children's use of phones while at school. When the kids come in, they leave their devices in one of two baskets sitting on the teacher's desk, a blue one for boys and a red one for girls. Rose, however, carries her mobile on her at school, kept on silent. After a bitter divorce, she has started a new relationship and is exchanging a lot of text messages with her new boyfriend. She admits, laughing, that sometimes she behaves like a teenager, checking her phone under the desk for incoming messages.

Alda

Alda worked in a government office in Italy where the rules were very strict. At the time, 2011, there was only one computer in the office, and only her boss was entitled to access it. If anyone had to send an e-mail to another department, they had to ask the boss for permission and the password. Mobile phones had to be switched off. So when Alda needed to call home or make a private call, she "went to the toilet." It's hardly an ideal place for a conversation, though: other people were there making calls as well, there was no privacy, and, of course, in the background there actually are toilets flushing.

Norma

Norma worked for a grand hotel where staff were not permitted access to their mobiles on the premises, so they were going outside if they wanted to make calls or check messages. Although the management tried to stop the practice, they still allowed smokers to use the backyard for smoking. So staff wanting to use their mobiles simply pretended to step out for a cigarette and made a call instead.

The Rules

The rules that different institutions have introduced to regulate the use of private communication devices are not readily available, except for the very explicit rules in schools and prisons. Use of mobile phones is banned in most schools, with the phones either switched off while on the premises or left in some prescribed place (locker, class basket, coat) and outlawed in prisons.[1]

In the workplace formal regulations about the use of cell phones, if present, are usually the responsibility of HR departments, alongside disciplinary rules concerning such matters as harassment, use of company equipment, and duration of breaks. Internet-related activities, however, tend to be monitored and controlled through IT departments, which can block access to some services. In 2010, for instance, in Switzerland a survey commissioned by the IT department of the federal administration showed that Facebook was the second-most visited site in the administrative offices of the federal government. Accordingly, the Conference of General Secretaries recommended that departments block access to Facebook. Six of them, along with the Federal Chancellery, decided to apply the measure This measure also was followed by the post office, several local authorities, and the bank UBS.

In France, labour law offers guidance on the use and abuse of an employer's equipment (including, therefore, companies' computers and phones), but little on the at-work use of private devices such as personal smart/mobile phones. As people have more access to the Internet through their own phones, it will be ever-less clear how companies will monitor, identify, and eventually sanction people who are engaged in private communication activities during working hours.

At present all French legislation is about communication using company-owned devices. An employer can examine the content of employee e-mails on a company account unless they are explicitly marked as private, and all company equipment put at the disposal of employees is for professional use only unless the employers, at their discretion, agree to accept some level of private use. La Commission Nationale de l'Informatique et des Libertés (CNIL) encourages a level of tolerance from employers so long as private use of the Internet is reasonable and does not affect the security and the productivity of the company.

Where the employee abuses this tolerance, the company is entitled to invoke the law, and there have been cases in which an employee has been fired for his digital conduct. La Chambre Sociale de la Cour de Cassation in France accepted the dismissal of an employee for abuse of Internet usage as a case of gross misconduct. In this particular case the employee had spent over forty hours in a period of less than a month carrying out personal activities over the Internet (or an average of two hours per day), with peaks at six hours on a single day. He was the only employee to have access to the computer, always took the time to erase the traces of his connections (cookies, temp files), and could not provide any evidence of any professional activity during that time.

In London in 2014 Maria Ramirez Angel was asked by one of her informants, who works as a cleaner for a company that does the maintenance of office buildings, to accompany her to a disciplinary hearing and serve as a translator. Her informant was being questioned and was on the verge of being fired because of excessive use of her personal smartphone. Unbeknown to the employee, the office building that she was cleaning every night had video surveillance cameras in most spaces. The employee had in recent months been filmed spending an excessive amount of time on the phone during working hours, and the managers of the company wanted to dismiss her on these grounds. As the employee was pregnant at the time, she explained that this was an exceptional situation, and she had been talking about her situation with her family and taking advantage of the calls to rest. In the end she was not laid off, in virtue of the laws protecting pregnant employees, but she was formally requested to stop using her phone during working hours. In private the employee told Maria Ramirez Angel that calling at night from work was the only time she could call her family in South America because of the time difference and her busy schedule.

The use of video surveillance for controlling the use of personal mobile devices is becoming more systematic in a range of work environments. Installed initially for safety reasons, they are now being more or less explicitly used as deterrents for potential abuses in the use of personal devices. In the case of the railway accident discussed in the next chapter, the board that investigated the disaster made the recommendation to install video cameras in all train cabins to monitor the use of phones by train drivers who find themselves alone in the cabin and cannot be surveilled.

The Productivity Argument

As the Cour de Cassation (the Chamber of the Supreme Court of Justice) in France mentions in its deliberation, besides issues of security, the main damage that private use of the Internet can cause at work is a significant reduction in productivity. In the case described above, the time the employee devoted to private activities was, by any standards, excessive. However, the reasons advocated to qualify the misconduct highlight the main argument against private communication in the professional domain: given the chance, employees will waste their working time on private communication channels, thus reducing their productivity and their companies' profitability.

The Economist (29 October 2009) quoted a study by an IT company specialising in systems for monitoring and filtering website access. The company, Morse, claimed that in the UK, employees were spending forty minutes per week of their working time on Twitter, Facebook, and other social media, and that this was costing UK businesses millions of pounds.

More recently the data-recording capabilities that are increasingly possible with most activities involving some transaction or interaction with digital systems are being used for a closer scrutiny of user activities. Logs of applications, transactions, and Internet activity are a normal procedure at work, but we are now also witnessing the use of geolocalisation, for instance, of company vehicles, supposedly in case of accidents or major problems. Video surveillance via CCTVs, as mentioned above, is also widespread in many work premises. Most employees are becoming aware that their actions are being systematically logged and checked and are, therefore, either using personal devices for some activities or elaborating particular strategies to maintain a sense of agency. Interestingly, surveillance is not always from above but can come from customers or other employees, as demonstrated by some famous YouTube videos of bus drivers chatting on the phone while driving. It is possible to find numerous photos on the web of workers, usually in some uniform or overalls, engrossed with their phones, accompanied by comments such as "This is how our tax money is being spent." These cases of "sousvellance" contribute to maintaining the norms regarding the division between private time and professional time.

It is clear that there is an element of Foucauldian disciplining in these practices. As jobs have evolved into "knowledge work," which require many hours of information manipulation, often on digital devices, discipline is not exerted anymore on bodies but on minds. Control must be exerted on attention because it is the human faculty that is being employed to carry out the job and it is that function being sold by the employee to the employer. Capital is being made on the exploitation of the attentional faculties necessary to transform and manipulate data and information. Controlling attention is, therefore, essential to ensure a good level of productivity.

What Exactly Is at Stake?

Year after year statistics show that the average mobile call in the United States as well as most European countries is two minutes long, a figure that has actually decreased over time as voice calls become shorter rather than longer. There are differing figures reported by different sources regarding the amount of time spent on social networks. Nielsen (2014) reports about seven hours per month on Facebook on mobile devices and six hours on PCs. If we look at the OFCOM data for 2014, they present the following daily (including home) usage figures: less than 10 minutes per day on mobile voice calls, 68 minutes on the Internet, 232 on TV, and 28 minutes using a mobile phone (a figure that includes a variety of activities on smartphones). Interestingly, for 2014 Nielsen reports longer time on mobile devices, up to 62 minutes per day, but this may include tablets, and only a third of this time is dedicated to communication and social media. How are these minutes distributed over the day? Time spent on personal devices/Internet use is unlikely to be clustered in one block, more likely to be dotted through the day in units of a few minutes. Nielsen (2014) reports that smartphone users access their phones on average nine times a day in the UK. This is not, therefore, a continuous process but rather an interspersed activity. We can derive from these statistics that mobile devices occupy something in the range of one hour per day, of which a third is dedicated to communication and that these activities are dotted over the day in periods of relatively few minutes. We will discuss later the cognitive impact of these minutes of communication in the flow of activity; here we can assess the impact of the time dedicated to personal communication during the work day.

The work day has changed over time, and there still are significant differences in labour legislation between countries and in practice between professions. Although in Europe historically we have seen a progressive decline in official working hours—from an average fifty-hour week to a thirty-eight-hour week—there is no doubt that there has been an increase in productivity during the same period. In fact, the reduction in working hours has been determined by an increase in productivity. According to the US Bureau of Labor Statistics, in 2009 business productivity in the United States grew by 8.1 percent, even as unemployment grew to over 9 percent and the country was in a recession.

Commentators such as Madeleine Bunting, in her book *Willing Slaves* (2004), argued, however, that the trend is reversing and that the last few years has seen an increase of pressure on workers. For instance, in the last seven years we have seen a significant rise in the number of employees working in excess of forty-eight hours a week, rising from 10 percent in the late 1990s to 26 percent now.

The number of people working a long week has also jumped. Estimates in 2000–2002 suggest that those clocking up to sixty hours a week have increased by a third, which equates to one-sixth of the UK labour force. In October 2010 a study by the French Ministry of Labour[2] showed a 14 percent increase in the number of supplementary hours (hueures supplementaires), which means about ten extra hours per quarter, with some professions totalling twenty-five extra hours per quarter per employee on average.

In essence, most rational modern economic models of productivity agree that it is impossible nowadays to isolate labour productivity as a unique independent causal variable. Economic models that directly equate labour time to productivity are ignoring the myriad other factors that affect work production. The equation is rendered even more complex in the knowledge-intensive activities that constitute many UK jobs. There is, for instance, a continuing debate in the social services (mentioned in the Munro review of child protection in 2009) about the balance between "real work," on the ground with people and administrative/bureaucratic tasks. Many process-control activities involved in the manufacturing of goods or energy require very little physical activity from the operators but high levels of monitoring. Work schedules often alternate long periods of low activity with peaks of very intense activity. The continuous Fordian cycle, modelled on an early-twentieth-century car production worker, is increasingly rare.

Rhythms of Work

Any expectation that a worker is engaged in her work tasks for the entire working day without interruption or distraction ignores both the cycles of work as they have been designed in most work environments and the natural cycles of attention of the human mind. We know that any number of external and internal factors can affect an individual's capacity to sustain focused attention (external distracters, internal emotional states, motivation). Management of attention is not just a question of willpower or control. Furthermore, as Kahneman described already in 1979, some tasks demand a greater effort of attention and have a greater demand in cognitive capacity than others that are processed in a more intuitive way. Most tasks, however, allow for people to go through cycles of focus, defocus, rest, and focus again.

A few very attention-intensive tasks—such as air-traffic control, where safety depends absolutely on the operator's undivided attention—involve very short work cycles (rest periods every ninety minutes, only ten days of work a month). Most other jobs tolerate moments of cognitive inattention, and people on the job are usually good at identifying natural break points for their activity cycles. How people use those break points has changed somewhat over time. (We shall look later at how they refocus after the breaks.) Stimulants such as cigarettes, coffee, and sugary snacks were often associated with breaks. Social interaction (chatting with colleagues) or moving around the office also created moments of relaxation.

But in the current context in which digital channels are available, it is probably the case that exchanges via text, e-mail, or voice channels are increasingly being used instead as stimulants or relaxation during short breaks.

Any notion of productivity must, therefore, incorporate some of the cognitive and social processes that organise work. The increase of overall productivity in the past hundred years did not correspond to a greater number of hours of work but rather to a better allocation of resources between, for instance, humans and machines, better management of resources, a higher level of education, a greater sense of autonomy and achievement in many roles, and so forth. To equate productivity to work time is both simplistic and wrong.

We have to take into account that the rhythms of work and attention change as jobs increasingly become knowledge and network intensive. Many of the schedules that are still being applied in most workplaces reflect the principles of effort management of a hundred years ago, when workers needed to rest from intense physical labour. Tea and lunch breaks rested the muscles and legs of a manual worker but may be unnecessary for a person who spends the working day sitting in front of a computer.

In the glory days of the Internet boom, technology start-up companies introduced pinball machines and Playstations, chill-out rooms, and gyms for their staff—compensation for the monastic life that young programmers were living, cloistered in their offices for days on end in search of fabulous innovation. However nerdish, these "toys" did capture the change in needs of this class of workers: young techies did not need to rest in front of a big meal; they needed to play and run in order to recapture focus and creativity.

It is likely, therefore, that the modern employee habit of shortening lunch periods and increasing the number of short breaks in which they surf the web or communicate is a better solution to combat the type of fatigue that data- or knowledge-intensive jobs produces.

The Social Nature of the Private/Public Divide

Considering the actual number of working minutes involved in private communication, the productivity argument looks, by and large, irrational and, I would argue, is more about the disruptive social nature of private communication than any effect on productivity. When employees receive or make a private call from their workplace, they are violating the separation of the private/professional spheres. They are challenging the historic principle of separation of realms.

In fact, when such calls, messages, and photos happen, many things are going on:

1. As Erving Goffman (1959) says, they are "showing the backstage," manifesting their private self by their intonation, choice of words, topics. It is a bit like seeing your colleague in pyjamas.

2. They are overtly breaking the social contract that binds the worker to give her full attention for the duration of the work time in order to receive compensation. Use of a mobile phone can be far more ostensive than simple daydreaming or chatting with colleagues, which makes lapses of attention very apparent.
3. They are violating the group principle that requires the worker to be focused on the people around them: they are engaged in remote relations from which colleagues are excluded.
4. They are breaking the illusion that the work environment is more important than the private one.

So although, objectively, each call, text, message, or picture represents a very minor loss of time, the social "damage" it is doing to the organisation is much greater, the more so when many individuals receive and make such calls. Taken together, they shatter the illusion that the professional environment is self-contained and fully separate from the private one. Professionals who are in charge of maintaining the organisational structure of a work environment or people who themselves prefer to keep the two realms very clearly separated find these intrusions of the private very difficult to handle. Different organisations will cope with this phenomenon in different ways, ranging from allowing the group to self-regulate behaviour, to demands for restraint, to explicit rules demanding phones to be left at the door.

E-mail, however—by far the most used channel of communication at work—does not give rise to any of the concerns mentioned above. The fact that it is written, usually carried out on the office computer, and embedded in other work activities, makes it less disruptive. The illusion of work and focus is not broken, so it is less controversial. The fact that, in terms of time consumption and productivity loss, it is probably far more significant than mobile phone use is tacitly tolerated. In our research in Switzerland we saw that the peak of private e-mail was during working hours.

Interestingly, as messaging and chat in its various forms (from Whatsapp to Wechat) have grown to be major private communication channels, they have also started to create the same negative responses as voice calls. Although messaging services are written channels and, therefore, should pass under the radar just like email, they seem to elicit more disapproval. This, in my opinion, is because they are perceived as semi-synchronous, and are thus equivalent to a conversation. Again, an employee engrossed in a distant conversation and relationship, albeit in writing, is perceived as more threatening to the social organisation than an asynchronous series of written emails. The sense of threat and violation of appropriate professional behaviour is not simply an issue of written vs. oral, silent vs. noisy; it has to do with excluding working partners from the personal sphere.

Social Class and Access to Private Communication

My contention that the issue of regulating private communication from the workplace is not a purely rational and economic principle but rather a highly charged social one is supported by the very clear differences between types of employees concerning permitted access to private communication. Nielsen Digital Consumer Report from February 2014 shows that half of employees do some social networking at work, but 57 percent of employees from higher-income households are networking from work, versus 36 percent of employees from lower-income households.

I have described several cases in which workers are highly restricted in their use of private devices, but there are many work environments in which even in 2010 there were no restrictions on the use of Internet or mobile phones as shown by the following cases recorded at the time.

Jenny is a designer in a web and advertisement agency. She has a big computer screen because she has to work with many programs at once to manipulate the designs she is creating. So her screen is full of overlapping windows that are constantly open. One of the programs open in the background is Facebook, and another one is her private e-mail. She switches to Facebook a couple of times an hour, just to see what is going on and occasionally comment on a picture or post. She also sends off a few e-mails every morning to reply to someone who has written to her. She says that she has stopped using instant messaging because she found that was really disruptive.

C and M are a couple who met and are now both working in the research lab of a pharmaceutical company. They work on different floors and often communicate by instant messaging—for instance, to decide when and where to go for lunch. Their department has no restrictions on access to the web, so they normally do their private e-mailing from the office. Neither of them is keen on social networking sites: they don't have a profile anywhere apart from LinkedIn, but they never look at it anyway. M says she receives the occasional phone call from her parents and friends but tends to keep calls short so as not to disturb her colleagues. When she is engaged in experiments in the lab, she switches her phone off because she can't be interrupted—and handling a phone is tricky with protective gloves on. She is puzzled by my questions on imposed restrictions—she can't imagine anyone in her department wanting to do that.

Mr R has been managing his own printing company for thirty-five years. He has fifteen employees working for him, a few in administration and the rest involved in the printing process. He is in charge of commercial relations as well as running the firm. He has just one mobile phone that he uses for all his calls: to clients and his secretary or to his wife and his friends. He is proud of being always reachable. Some of Mr R's long-standing major clients have become friends or at least friendly, and he may call them to arrange a lunch that has no real business objective. Similarly, his e-mail is a mix of purely professional and semiprofessional messages with his vast circle of connections. He considers that his success in the printing business stems from the strong personal ties he has been able to build over time. The only employee in his company with a company phone is a driver who is in charge of deliveries.

In working environments that offer a high level of autonomy and self-direction to the workforce, the integration of private communication has been seamless. Research units, self-employed professions, and organisations based on goal or project evaluation have incorporated personal devices without difficulty because they were already set up to trust their employees. This level of trust, however, has in general been associated with jobs of a higher status, carried out by people with higher education.

In the days when the only communication tool available at work was the landline, any interpenetration of the private and professional was the privilege of the elites—senior professionals, managers, and academics—or was earned by moving up the ranks:[3] with the corner office also came a personal outside line. This was a badge of status granted to people who were seen as able to exert self-restraint and had proved their commitment to the institution they belonged to: crudely, people who could be trusted to get their priorities right.

Decades later things do not seem radically different. Workplaces that are structured with systems of supervision and time-based productivity remain more or less rigid in excluding the new communication channels.

In our interviews regarding use of communication devices while at work, across several European countries we have encountered a very clear distinction in attitudes between those workers who assume they are entitled to unlimited access to private communications channels and those who are cautious and circumspect about using anything because they fear sanctions. It is no surprise to find a considerable difference in the type of jobs these attitudes reflect. Creative agencies, consultancies, and research departments, for example, impose no limits on access to Internet and other devices; distribution, transport, manufacturing, and catering firms mostly do. Even within the same companies there are vast differences in attitudes and access according to rank and role.

Essentially the greater the autonomy granted to people in managing and executing their work, the more likely it is that they will be entitled to self-regulate their use of digital communication devices. The issue at stake, therefore, is not the use of Internet as such but the level of trust and control that is exerted within an organisation. I would go as far as to argue that looking at the attitudes and practices regarding the use of communication channels is a good predictor of the type of management of an organisation—whether, in other words, it is structured on goal achievement and self-organisation or on hierarchy and task execution.

For these same reasons, people who work from home—either because of the nature of their profession or because their employer requires it—are regarded with some suspicion, and systems of control are applied to regulate their commitment. Freelancers who work from home—from journalists to clothing manufacturer pieceworkers—tend to be paid a fixed fee for a piece produced, not for the hours taken. This means they carry the responsibility themselves for using their time effectively.

There are two solutions to control professional commitment and production: either enforce some form of regulation to separate the working activities from the rest of the workers' lives, thus depriving them from the potential disruption of other personal events, or force individuals to impose the self-discipline to dedicate their full attention to the task by linking their revenue to the use of their time.

Cross-Cultural Bias: Private Communication in Developing Countries

The historical and cultural nature of the division of realms is also made evident by cross-cultural studies. Alternative models of mixing the private and personal sphere in professional environments are seen in regions where the micro-entrepreneur model is among the most successful economic solutions. Studies of micro-entrepreneurs in West Africa, for instance, show that the economic system is dependent on very close collaboration and communication among kinfolk. In a report by the International Labour Organization on female entrepreneurs in Africa (Finnegan, Howarth, and Richardson 2004), the researchers concluded that although for these women a family and dependents at home was occasionally a hindrance in the development of a small business, generally the family was experienced as an asset.

One typical case study described a fisherman and his tiny crew, his wife selling the fish at the market, his brother transporting the fish, and another relative buying the fish for a restaurant. Similar close-knit networks of collaboration have been described for taxi drivers, coiffeurs, and shop owners.

"Mobile for development" initiatives, or M4D, have been financing the acquisition of mobile phones in developing regions, with the aim of supporting micro-entrepreneurs in their daily business. However, just as in Europe, a majority of the calls made with these mobile phones are of a private nature. Jonathan Donner's study of 277 micro-entrepreneurs in Rwanda (2006) noted that only one-third of the calls they made were business related.

Recently Donner (2009) has written a paper trying to quell the M4D organisations' concern that the phones they were subsidising were being used for idle chit-chat and private matters. He points out the importance of "blurring life and livelihood."

> Put another way, though people might be willing to adopt a mobile and use it in ways that eventually will be beneficial to them in a "development" sense, the reasons many of them will do so will likely have little to do with these development outcomes. If mobiles were not so enjoyable, fewer people among the poorer half of the world would be willing to purchase them, and putting mobiles in the hands of current nonusers

for developmental purposes would be a more difficult and expensive proposition (Donner 2009: 96).

Concerns in the development aid communities are very similar to those we have seen in corporate Europe and the United States. Agencies that subsidise or support the distribution of mobile phones for economic development frown upon the intense personal use that is being made of them, for exactly the same reasons corporate Europe frowns upon employees' communications habits. The cultural root is the same. However, mobile communication has been systematically hailed by economists as the developing world's means to expand their range of services and cut out intermediaries, ignoring the fact that the success of micro-entrepreneurs relies heavily on the tight link of connections they entertain that can be better sustained by better communication.

Gender and Private Communication

The lack of tolerance in the workplace for private conversations reveals another profoundly cultural set of beliefs on the nature of work. Feminist sociologists such as Valery Bryson (2007) have commented on the exclusion of private care, such as that provided by women to their children, elders, and spouses, from the category of work. In her argument parental care is a type of work that is principally carried out by women and that is undertaken either in addition to professional work or instead of it. In any case, because it is not recognised as "proper" work, it is not compensated or regulated.

Following her argument, the fact that we consider as "private" all calls or messages sent by working adults to the people they are taking care of reflects this cultural bias. Checking that the children are home or that an elderly relative is safe is a "personal" objective inasmuch society currently does not recognise care giving as a job of social relevance. The oppositional nature of care-private versus work-public renders all care-related exchanges from the workplace as violations. If care was socially considered a "job," a call home would be equated to a call to a client or colleague.

Schools and Private Communication

Currently most European countries include information and communication technology (ICT) in the school curriculum. In France, for instance, ICT is taught from primary school onwards. ICT was introduced as a compulsory subject relatively soon after personal computers became available so as to ensure that pupils acquired a basic level of mastery of the tools seen to be essential for any future professional role. Computer literacy courses at school also avoided a digital divide between children who had the means to own a computer at home and those who did not, as well as recognising that the predigital generation of parents might not be able to teach their children at home. With computers becoming much cheaper and widely adopted in homes as well as workplaces, access at school to computers has become much less crucial. Already by 2008, 79 percent of children in Europe[4] had regular access to computers at home. However, the approach to ICT teaching continues to be influenced by the professional metaphor. On the one hand, school computing classes provide a basic familiarity with the major software programs available and, increasingly, with computer programming languages. On the other hand, socially complex issues such as behaving in social media environments, online identity, privacy, personal disclosure, or management of attention is generally not an integral part of computing classes but is occasionally included in disciplinary assemblies. Many schools are happy to encourage pupils to learn to code, write blogs and online journals, or discover educational resources and games. The more private realms of social networks and communication channels, however, are seen as beyond the remit of the school.

This attitude in educational establishments, which from a practical point of view is fully understandable, perpetrates the discourse about the unacceptable presence of private communication in institutional environments. Schools initiate the children to a practice of concealment when it comes to personal devices, making it very clear what it considers to be acceptable productive digital activities. By formally restricting and punishing the use of personal devices, these schools are enacting exactly the same strategies as the workplaces that will employ these children in the future. Schools are signalling that children cannot be trusted and are thus taught how to manage their attention, and they are reinforcing the view that some people do not have the capacity to focus and to be autonomous in their relation to technology. Schools are presuming, exactly as the most restrictive work environments we have encountered, that students are incapable of establishing appropriate priorities. By entertaining and enforcing these beliefs, schools are renouncing the opportunity to help students learn how to become autonomous in their attention management and use the tools as learning devices.

An Early Case of Successful Integration of Private Communication

The little town of Serris, just thirty minutes to the east of Paris, is part of the Val d'Europe project started after the signing of a contract between EuroDisney and the French State.

EuroDisney (which was opening a theme park in the area) undertook to sustain local economic development and has participated in the urban development and the transformation of Serris from a rural village of nine hundred inhabitants in 1995 to a commercial town with nine thousand inhabitants in 2010. As the town extended, so did the services that the local administration had to offer: new schools and childcare services, new streets to be maintained, new transport facilities, new parks, new spaces and equipment for sports, cultural activities, new financial and administrative services. The local administration, therefore, grew exponentially, from 20 employees in one small office in 1998 to 220 employees in 2011. From a small, tightly knit group of colleagues who had worked together for years and for whom their town hall was like their home, the organisation became complex and diverse. Gardeners and teachers coexist with financial officers and technicians. A new large and modern Mairie (Town Hall) was built to host the office workers and the elected members of the town hall.

When the new director of services (Directeur General des Services) Didier Vaubaillon was appointed in 2006, he saw that the development of the organisation had been "enthusiastic but unstructured" and that the moment had come to professionalise processes and roles. He was adamant, though, that this should not compromise the human values that he believes in and that he felt were animating the employees of the town administration.

A plan was made to reorganise the services, breaking down barriers and a lack of communication between people and departments and generating a flow of information that significantly increased the cooperation and, ultimately, the efficiency of the services. The main theme of the change was trust. Mr Vaubaillon supported people individually by creating an environment that tolerated errors and promoted learning, by believing and convincing everyone that incompetence could always be transformed into competence. He also recognised that people cannot give of their best until they have a clear understanding of what they are doing.

The outcome of this process? Very high levels of satisfaction among the population of Serris; increasing property values, in spite of the strain on services, as more and more people want to move to Serris and none wish to leave; one of the highest levels in the country of services for preschool children; and an efficient system within the Mairie, which means, for instance, that a new resident with lots of official paperwork to complete will be helped through the whole process by one dedicated administrator. Within the organisation itself, procedures have broken down the walls between services and shared problems are solved jointly. Job satisfaction is very high, and relationships between colleagues are extremely friendly and supportive.

We asked whether there were any restrictions on private communication at work. Mr Vaubaillon was amazed by the question. "Of course not," he said. "Why on earth would we need that?" All employees were free to use their own mobiles whenever they liked. Managers were given a phone as part of their job and were free to use it for private calls as well (extending the organisation's prepaid minutes at their own expense). Employees can access their own e-mail and social network site on the organisation's computers if they wish. The only restriction they had to impose was on MSN because the IT people realised it was a vehicle for viruses.

Mr Vaubaillon believed that people are themselves the best judges of whether their private communication would interfere with their work. They had to intervene only twice: once when someone was downloading films and another time when they noticed their landline bills showing some lengthy calls abroad. Both cases were handled individually, the people involved were asked to stop, and this did not give rise to general restrictive policies.

It is clear that people were behaving responsibly and that problematic behaviour can be discussed between the people involved without having to impose regulation. Rather than restricting access to the Internet, Mr Vaubaillon and Mr Legasa, his deputy, seemed to be far more concerned about ensuring that everyone had access, including employees working outside the office.

The attitude at the Mairie de Serris towards the use of private devices confirmed for me once again that the issues lying behind restrictions are organisational and social. In Serris the challenge of the years of expansion of the town hall had been to build trust, autonomy, and respect in a growing organisation, and this meant that individuals had to be recognised as full and responsible persons. Reconciling the private and professional spheres was an obvious and inevitable step—too obvious, in fact, to even need to be discussed. When everybody knows what they are doing and why and there is a culture of achievement, the issue of attention management becomes obsolete.

We may consider that the environment in a small public administration is privileged in many ways: there is the security of employment guaranteed by being a public servant, employees are selected carefully, and many of them have gone through higher education. However, it is also an institution that has, by definition, embodied the values of bureaucracy, of an unfaltering distinction between private and public interests. It has inherited military models of hierarchy and authority and has prided itself on its rational exclusion of personal pursuits. The smooth inclusion of the intimate sphere of the employees by 2011, in such a space, was truly a radical social transformation.

Chapter 5

Accidents, Distraction, and Private Communication

Controversy surrounding the use of private communication devices is not only between employers and employees or teachers and students; it is also rife in the general public and indicative of how society as a whole is learning to manage these innovations. There is a growing concern that the addictive nature of digital devices gradually damages people's capacity to assess where and when it is risky or inappropriate to use them. There is a parallel fear that these devices modify our cognitive capacities, reducing our ability to concentrate and to process information.

In the last few years there has been considerable media coverage of accidents involving trains, buses, or planes in which operators in charge of the safety of their passengers were engaged in private communication. These incidents have given rise to extensive investigations. For the social scientist these accident reports offer minute-by-minute descriptions of a person's tasks and activities and provide exceptionally clear data on the use of communication devices during work. The very detailed analyses of the sequences of actions that have led to the accidents provide unique insight into the way personal conversations are being inserted into the flow of professional tasks and help us understand the nature of the threat posed by inappropriate use of personal communication devices.

These cases highlight the particular social opprobrium provoked by people who attend to their private affairs on duty, and they have triggered the enforcement of new restrictions on mobile phones in various environments. The collective shock provoked by fatal accidents often gives way to regulations and new laws that would have otherwise been debated at length. Calls to ban personal communication devices

Intimacy at Work: How Digital Media Bring Private Life to the Workplace, by Stefana Broadbent, 79–89.

in one or another environment usually emerge in the aftermaths of these tragedies. The shock effect, as Naomi Klein (2009) states so clearly in her book *The Shock Doctrine,* leads to greater social control. The Chatsworth collision that we examine below, for instance, not only led to a number of new regulations in the railway industry but also sped up new "texting while driving" legislation in California.

A Train Accident

On 12 September 2008 a Metrolink commuter train in Chatsworth, California, failed to stop at a red light and crashed into a freight train. Twenty-five people, including the train engineer who was driving the commuter train, were killed. This was the worst train accident of the decade in the United States and received huge media attention. In that context the finding that the train engineer had been text messaging a few seconds before the crash pushed the authorities in California to immediately ban mobile phones for train crews (and then introduce legislation against texting while driving cars).

Mr Sanchez, the train engineer, had been texting all day on duty and had a history of using his cell phone at work. He also had been inviting young train enthusiasts into the train cab (the train "cockpit") and had even let some of them operate the train. He was certainly contravening a number of safety regulations on a regular basis and breaching standard security procedures. However, his behaviour was also set in a highly risky context of single-track railway lines and lack of automatic breaking systems on trains. It is worth spending a few paragraphs on the analysis of this accident because of the extreme nature of the engineer's communication behaviours and the regulatory consequences it brought about.

Description of the Chatsworth Train Collision

On the day of the accident the Metrolink commuter train and Union Pacific freight train were travelling in opposite directions on the section of single track that runs between Chatsworth station and Simi Valley through the Santa Susana Pass, where three tunnels under the pass are only wide enough to support a single track. The line's railway signalling system is designed to ensure that trains wait on the double track section while a train is proceeding in the opposite direction on the single track.

The Metrolink train would normally wait in Chatsworth station for the daily Union Pacific freight train to pass before proceeding unless the freight train was already waiting for it at Chatsworth. Tests of the railway signal system after the accident showed that it was working correctly and would have shown proper signal indications to the Metrolink train, with two yellow signals as the train approached Chatsworth station and a red signal at the switch north of the station. That day, however, Mr Sanchez apparently did not see the red light and continued on his track.

A few days after the accident two teenagers reported that they had received text messages from the train engineer just minutes before the crash, and it emerged that the engineer was using his mobile phone on duty. Further investigation showed that Mr Sanchez had sent a text message only twenty-two seconds before the crash: he had presumably been engaged in texting and had failed to see the red signal. Analysis by the National Transport Safety Board (NTSB 2010a), in charge of the investigation, of Mr Sanchez's phone communications found that he had sent and received fifty-seven text messages that day. The content of the messages also revealed that he was planning to let a teenage train enthusiast come and operate the train, as he had apparently done in the past. It emerged that Mr Sanchez's employers had already warned him on two occasions about his intense use of his mobile in the cab.

In 2003, following a fatal head-on train collision in Texas in 2002, the NTSB had already recommended that the Federal Railroad Administration introduce some regulation regarding the use of mobile phones by train crews. However, at the time of the Chatsworth accident there was still no federal regulation. The day after the NTSB confirmed that the engineer had been texting and less than one week after the accident, the California Public Utilities Commission passed an emergency order to ban the use of cellular communication devices by train crew members. A week later texting while driving a car was likewise outlawed in California. The emergency order stated that all personal devices and peripheral devices must be turned off and stowed out of sight at all times when crew were required to be performing service. Only when a crew member was on a break or in the crew room could he use his personal devices.

Analysing This Case

This terrible accident is an extreme case in many ways because the engineer breached so many basic principles of safety and professionalism. However, as in all accidents, multiple background causes and factors created the conditions within which specific events or actions then triggered the disaster (Hollnagel, Woods and Levenson 2006).

- According to the NTSB, Southern California has more tracks shared by freight and commuter trains than anywhere else in the country, so it has a greater need than most for collision avoidance systems, in use in parts of the Northeast and between Chicago and Detroit, which automatically brake and stop the train.
- The train engineer, Mr Sanchez, was on split shift: he worked an eleven-hour day composed of two shifts with a four-hour break in the middle. That day he had worked from 6 to 10 a.m., had a break until 2 p.m., and started his afternoon shift that would last until 9 p.m. This type of shift has been recognised as allowing little rest to the drivers and reducing their level of vigilance and awareness.

- Remarkably, on the other train—the Union Pacific freight train—the train conductor had exchanged forty-two text messages that day while on duty. This suggests that Mr Sanchez's texting was not as extreme as it seemed. The detail of the investigation allows us to lift a veil and look into the reality of the job of train crews and consider what the use of personal communication devices reveals about that demands of this activity.

The fact that in the Chatsworth accident crew on both trains had been texting during the day suggests that this may be a common practice, which in turn means that the job is somehow conducive to this type of behaviour. As in other jobs, when tasks are too repetitive and boring, mental underload can push operators to look for outside stimuli. Isolation can drive operators to search for contact; excessive confidence can tempt operators to think they can successfully divide their attention. Mr Sanchez may have been completely distracted by his communication activities and simply ignored all signals, or maybe he was relying on routine events (e.g., the freight train always waits for the Metrolink train to pass first, so no need to check) to make vigilance unnecessary.

The train accident in Santiago de Compostela Spain in 2013 (Comision de investigation de accidents ferroviarios 2013), in which a high-speed train ALVIA derailed because it entered into a curve at a speed of 190 kilometres-per-hour when the maximum speed should have been 80 kilometres-per-hour, is a similar case to the Chatsworth accident. Here as well the train driver was engaged in a call just moments before the derailment. He had been called on his mobile by the train conductor to discuss the possibility of changing track in the next station in order to facilitate the descent of a family with a wheel chair. The conversation was found to have distracted the driver, who did not see the signals informing him to reduce speed before the bend. The train did not have an automatic alert-and-breakage system, and the track was known to be dangerous at that point. The driver survived, although seventy-nine passengers died, and he has been found guilty of homicide by professional recklessness and held responsible for his lack of attention. Following the accident, RENFE installed the Automatic Breaking and Announcement System on the track a few kilometres before the Santiago station, thus ensuring the automatic breaking of the trains. Interestingly, the official report of the accident has in and appendix an information leaflet regarding the use of mobile devices, circulated among the RENFE personnel, with images of Chatsworth and the regulations that were introduced as a consequence of the accident.

An Air Accident

On 8 August 2009 a small aircraft and a helicopter collided over the Hudson River, and all nine people on board both were killed. In the following days it emerged that the air traffic controller, who supposedly was following the small plane, had been chatting on the phone with a friend just before the collision. Transcripts of his joking conversations about dead cats and barbecues were published in many newspapers, and the controller and his supervisor were put on leave. The investigation showed that the controller had not followed the procedure correctly by not transferring the aircraft to the Newark control centre in a timely fashion. He did not inform the pilot of potential conflicts, although he couldn't have warned the pilot about the helicopter because that aircraft was not visible on his radar.

As in the case of the Metrolink train crash, the context of the accident was highly risky, and as some experts declared in the following days, "an accident was waiting to happen." The area over the Hudson where the aircraft were flying is an unregulated area where pilots fly free of air traffic control. Every day there are more than 250 unregulated aircraft flying in that corridor, and the National Transport Safety Board had already called for a better regulation of the area.

The Hudson River Air Collision

The accident took place in an area known as the "Hudson River Visual Flight Rules Corridor," which extends from the surface of the river to altitudes of eight hundred to fifteen hundred feet at various locations along the Hudson River in the immediate area of New York City. Within this corridor aircraft operate under visual flight rules, according to which the responsibility to see and avoid other air traffic lies with the individual pilots rather than with the air traffic controller. Many small aircraft that need to transit the New York metro area use this corridor to avoid longer circuits that bypass the restricted areas dedicated to commercial flights. The corridor is also heavily used by helicopter tour companies, which take passengers on sight-seeing tours of New York City.

The helicopter, carrying five Italian tourists and its pilot, took off from the West 30th Heliport at 11:52 a.m. At about the same time, Teterboro Tower radioed the Piper plane at take-off, requesting him to pick his flight path towards Ocean City, and the pilot decided to head there via the Hudson River. The aircraft was then instructed to contact Newark airport. Soon after, a controller at Newark who was concerned about aircraft in the plane's path called the Teterboro controller and asked him to attempt to reestablish contact. Neither controllers managed to contact the plane and change its heading. At that point a radar alert about a possible collision went off in both the Newark and Teterboro Towers; however, the two controllers did not remember seeing or hearing them. While heading south, the plane was seen to be behind the sightseeing helicopter, which was going about half as fast. The pilot of another helicopter saw the impending accident and attempted to warn both the helicopter and the plane by radio but received no response. The wing of the Piper plane hit the helicopter, and both aircraft plunged into the river.

On 14 September 2010 the NTSB released the final report regarding the incident. The report discusses several aspects of the collision, including locations of origin of the aircraft, planned destination, air traffic control communications, pilots' training, aircraft detection equipment, and air traffic procedures and regulations. A number of factors are examined, and multiple causes, as always in the case of accidents (Perrow 1984) are identified in the report. The dense and complex airspace around the busy airports of New York, rules regarding the separation between aircraft in that zone, and a number of misunderstandings between pilot and the air traffic controller are some of the principal causes identified, as was the inherent limitations of the see-and-avoid concept operating over the Hudson. Although there was a good visibility, the airplane pilot may not have seen the helicopter because it was against the background of the complex building skyline, which made it difficult for the airplane pilot to see the helicopter until seconds before the collision described.

There was then a second probable cause: the misunderstanding relative to advisories. Just before take-off the pilot had been asked by the controller to select one of two routes, and he chose the route on the Hudson, which would have required him to monitor his own advisory system. Advisories are information about other relevant traffic in the flight-plan area of the aircraft. However, the pilot had previously requested to be given advisory information by air traffic control and was probably still expecting to receive this information from the controller. The Teterboro Airport local controller, however, did not provide continual traffic advisories to the airplane pilot, which he could have done, given that his workload was light; such advisories would have heightened the pilot's awareness of traffic over the Hudson River. In absence of any information from the controller in Teterboro, the airplane pilot may have believed there were no other potential traffic conflicts.

In addition, the report underlined that the Teterboro controller was making a private phone call and did not transfer the flight correctly to the next control area of responsibility, the Newark control. He did the automatic transfer via the system interface but did not conduct the oral call to transfer the airplane from one frequency to another. The air traffic controller, in the minutes before the crash and just after handing off the Piper to Newark, had in fact been on the phone with a colleague at the Teterboro Operations, joking about barbecuing a dead cat that had been found on the parking lot. He had been using an internal line to call her and had been resuming a conversation that had started some minutes earlier before he handled the Piper.

The Teterboro Airport local controller unnecessarily delayed transferring communications to Newark Liberty International Airport (EWR), which prevented the EWR controller from turning the airplane away from Hudson River traffic to move it into less congested airspace. Not only did he not complete the transfer early enough, but when he did, he did not correct the pilot when he read back the frequency to contact the EWR controller, which meant that the pilot was on the wrong frequency and could not hear the warnings from EWR.

The NTSB (2010b) writes in the report that:

> The Teterboro Airport local controller did not correct the airplane pilot's read back of the Newark Liberty International Airport tower frequency because of the controller's non-pertinent telephone conversation and other transmissions that were occurring.
>
> The airplane pilot's incorrect frequency selection, along with the Teterboro Airport controller's failure to correct the read back, prevented the Newark Liberty International Airport controller from issuing instructions to the airplane pilot to climb and turn away from traffic.

The National Air Traffic Controllers' Union then issued their own press release, disputing some of the statements in the NTSB's report. Consequently, the NTSB retracted some of its statements regarding the controller's part in the crash, saying that the controller could not have warned the plane about the helicopter because the tour helicopter was not, in fact, visible on the controller's radar. Regardless, the Federal Aviation Authority put the controller and his supervisor on leave. The supervisor had left his position for thirty-five minutes and was absent during these events. The NTSB claimed he should have been there to warn and discourage the private phone call.

Analysis of the Accident

This case, regarding the causes and the responsibilities of the different actors in the event, pinpointed the air traffic controller as having contributed to the overall catastrophic sequence of events. The phone conversation of the controller did not seem to be identified as the primary cause, but it did reduce his level of vigilance and communication with the pilot after he had "handed the plane off" to the following control centre.

In our view this case exemplifies one of the mechanisms associated with personal phone calls from the workplace: calls or other forms of private contact are often made in down times when there is a feeling that a task has been achieved. There is a cycle of activity, and when there is a sense that there is a break in the cycle, either because the task is completed or because the person has to wait for the next cycle, an opportunity for a conversation then emerges.

In this case the controller probably had a sense of closure of the activity regarding the Piper plane—it had been "handed off"—and did not have anything else urgent on his screen, so he resumed his call. He dropped the conversation as soon as he received a call from the Newark control centre. The question of his responsibility, therefore, lies in whether he "closed" his activity too soon—whether, in other words, he should have followed the plane longer or ensured that contact had been made with Newark or informed him better. The question is: Did he hurry the transfer because he wanted to resume the conversation with his friend, with whom obviously he shouldn't be conversing in the first place?

In the derailment of Santiago de Compostela, the train conductor had called the driver and engaged him in a conversation that lasted more than a minute and a half on an issue that regarded a station more than fifty minutes away (Ferrol was the second stop after Santiago de Compostela). When asked during the inquiry why he had not waited until the train was stopped in Santiago to discuss with the driver an issue regarding a later stop, the controller said that he was in a quiet moment; he would not have time during the halt in Santiago, as he had to monitor passengers and doors and knew that there was no GSM coverage after that along the route. He thus used that moment because it was a period without interruptions. When asked why they were using their mobile phones for this exchange, they both replied that it was the easiest means of maintaining contact.

Is Private Communication During Work Dangerous?

In all these accidents one of the people responsible for the security of the passengers was engaged in a private communication at the time of the crash. Such accidents inevitably raise questions about whether there is something intrinsically dangerous in letting people have access to their personal devices at work. Is their use so addictive or compelling that it can lead people to fail in their responsibilities? Is there a difference between keeping a smartphone in your pocket and having a daily newspaper folded next to the driver seat or a colleague who drops in to say hello?

Experiments that have measured the attention demand of cell phone conversations on drivers show that there is a huge drain of attention away from the main task. A well-known study showed that phoning while driving is equivalent to driving drunk (Strayer 2006). These studies, however, presuppose that a person is trying to multitask—driving and calling or texting at the same time. I would argue that in work situations activity is rarely continuous and that private communication is generally carried out in the moments between activities. It is more often a case of rapid switches of activity than multitasking.

The air traffic controller may have fallen into a typical mechanism associated with personal phone calls from the workplace: calls or other forms of private contact are undertaken in what feel like breaks in the cycle of activity, when a task is completed or when the person has to wait for the next cycle of activity to begin.

We can make the hypothesis that the train engineer driving the Metrolink also erroneously had a sense of having done the necessary activities involved in leaving the station and starting a six-minute uninterrupted trail to the next station. Something may have given him the impression that all was clear and that he could relax his attention and resume a text conversation. He may have thought that all was clear ahead and that he had ample time to send a few messages before he had to focus again.

Using personal communication as break fillers, small rewards for tasks accomplished, or ways to counteract the dullness of an activity is very common. People don't just drop what they are doing to suddenly send a message or an e-mail but instead wait for one of these moments (unless they have to interrupt what they are doing to answer a call, but in that case they are not the initiators of the action). The sense of closure or waiting for the next task depends on the nature of the activity as well as one's personal disposition. Different activities have different cycles, and therefore, down moments or closures can have different frequencies. Clearly there are also limitations in people's attention cycles.

When writing we may feel that completing a page or paragraph is an achievement that closes a cycle of concentration and can be rewarded by a small break of attention: going to get a cup of coffee, reading e-mail or the news, checking social media, or sending a message. Similarly, with manual work there are points of closure before a new set of actions start (e.g., waiting for a surface to dry before polishing it) that can also be used as breaks. There are also many activities that are highly predictable, and experienced workers know that at certain points there will be some prolonged periods of waiting. For instance, in manufacturing, expert machine operators often know exactly how long it will take for a process they initiate to have an effect. Until it does so, there may be nothing for them to do. These are typically moments in which a self-gratifying activity may take place—until recently perhaps having a cigarette, now maybe checking social media. The success of asynchronous channels such as texts, e-mail, or social media can be explained also by these patterns of work and attention. Using an asynchronous channel means that people can communicate at their own rhythm—when they have entered one of these down moments in their activity cycle—without interrupting another person's activity cycle.

Accidents may occur when operators mistakenly feel that they have finished a task or cycle and are waiting for the next event to act upon or when they have miscalculated the latency between the two cycles. In both accidents we've looked at the operators may well have released their attention too early, estimating that their cycle was done and that they had time for a short switch of activity before refocusing.

So we are back at the question of the potential danger of personal communication devices as sources of distraction. Does the presence of such devices in itself lower attention? Do they cause operators to rush their tasks in order to be able to switch back to their private activities? Do they eat up precious mental and attentional resources that operators should be dedicating uniquely to their tasks? Are operators becoming overconfident about their capacity to switch between work and personal activities?

As our research showed, the intensity of the relationships that are managed with mobile phones in particular and with digital media in general means that this form of communication is extremely compelling. It is also clear that these exchanges bring immense gratification that enriches and enlivens the ordinary working day. Most importantly, digital devices modify how we experience down time. Digital devices can fill those empty moments very effectively with a variety of very satisfying and rewarding activities, be they communication, music, or games. So the question shifts to one's time management of these personal activities and whether they engage users for spells of time that are incompatible with their monitoring/work activities. The risk is that operators miscalculate the time elapsing between two events that need their attention, prolong the pause between one task and the next, and miss vital indications that something critical is happening in their environment that requires their intervention. In the past, operators could anticipate the duration of a lull in activity and fill that up with another time-limited activity, such as a short exchange with a colleague, a coffee, or cigarette, and, therefore, match their pause activities with the attentional requirements of the job. The digital devices they are engaging with now may be affecting their sense of duration or their ability to match the duration of the pause in the activity with the duration of the digital or communicational activity. This difficulty of evaluating the time spent in a digital activity is often reported by users. People often talk of digital devices as "time sucks."

In terms of the human factor (Hoc 2001; Hollnagel, Woods, and Leveson 2006) perspective, which I wish to adopt here, the presence of a personal phone is particularly dangerous when the whole context is highly risky. When operators are alone, tired, or bored, when they have little to do for long periods and little time to recover from errors or external events, then undoubtedly a personal communication device is a threat. When the environment is one that breeds overconfidence and breaches of security measures (e.g., not wearing safety gear, making shortcuts in the procedures, skipping reporting, working more hours than allowed, etc.), operators will be more likely to feel that their task deserves less attention. But the threat is the outcome of a progressive isolation and automation of many tasks that involve the operation of machines. People have been slowly but surely relegated to monitoring tasks where they are expected to maintain a high level of vigilance with low levels of activity. They are also expected to be the last link, which means they need to intervene very quickly in the event of a crisis or problem. Tasks of this sort are encountered not only in transport but also in manufacturing and process control of utilities.

Any expert in human factors would question why so many human lives should rely on such fragile processes. How is it possible that a lapse of attention of one individual for such a short length of time can cause so many deaths? The context of both accidents—a one-track rail system and a densely trafficked unregulated airspace—creates the highly risky conditions in which diminished attention becomes fatal. Human error thus becomes the spark in an explosive environment where a multiplicity of risky factors are already combined, waiting for the last triggering element to occur.

There is a terrifying combination of isolation, automation, deskilling, flexibility, and insecurity—which Sennet (1998) would call the "corrosion of character"—that can cause people to lower their attention. When environments become poor social structures, eliminating the spaces for exchange, solidarity, continuity, and learning, then we can expect people to look for their gratifications somewhere else. The arrival and inevitable spread of smartphones and other personal devices that enable a variety of communication channels, information retrieval, and entertainment is bound to put a serious strain on professions that rely heavily on operators' attention. It is still relatively easy to regulate the presence of personal devices on duty because their appearance in the workplace is so recent. (Issues such as operator tiredness have been left unregulated[1] for many years because they have a far longer and more complex history.) But as digital devices become more and more entrenched in all aspects of daily activity and personal identity (Hutchins 1995), it will become increasingly difficult to simply ban devices from workplaces or other environments. As a society we have to address the issue of attention, analysing it as a process that we cannot simply regulate by imposition but must learn to handle in a more structured, socially sophisticated way. When environments carry their security in their processes and culture, individual lapses of focus can always be recovered.

Chapter 6

Conclusions: Communication and Attention

Throughout this book the notion of attention has been intertwined with communication. Chapter 1 showed that negotiating who receives and who gives attention is a fundamental criterion in the choice of communication channel. In Chapters 3 and 4 I suggested that controlling attention is the recurrent reason for limiting or banning access to digital communication channels: ensuring attention to other group members and to shared goals is perceived as incompatible with the use of private communication tools. Permission to manage one's own attention and communication channels is bestowed only on those who have proven their allegiance to the institution or their high-level position within it. Loss of attention is often a principal cause of major accidents involving the use of personal communication devices, as we saw in Chapter 5.

Why are these two processes so tightly linked? Why is it virtually impossible to talk of the social mechanisms organizing access to and usage of new communication devices without discussing the role of attention? I believe that in order to understand the underlying reasons for the social tensions the new communication devices bring with them, we must attempt to explain the strong relation between these two processes. The hypothesis of a human cognitive specialisation for sharing attention has been extremely popular in psychology, philosophy, and cognitive sciences for the last twenty years. And the evidence for such a skill comes from studies on primates, child development, and psycholinguistics.

Michael Tomasello, a psychologist who has worked for many years on the origin of language, explains that human language is unique not so much in the capacity

Intimacy at Work: How Digital Media Bring Private Life to the Workplace, by Stefana Broadbent, 91–95.

to vocalise in a flexible way and to create symbolic relations but rather in its fundamentally cooperative nature. The main precondition for the evolution of language, both philo- and onto-genetically, is the emergence of shared intentionality. Unlike other species—even the ones closest to us, such as apes and chimpanzees—humans are able to create shared goals. Although other primates can cooperate on individual goals, like hunting together to satisfy each individual's goal of eating, humans can devise objectives that are not individual but actually shared, such as growing crops, building a cathedral, or educating children. Humans have evolved a sense of collaboration, implying the capacity to imagine objectives that make sense for more than just the individual self.

Tomasello and colleagues (2005) suggest that beyond just understanding others as intentional agents and responding to them humans also understand others as potential cooperative agents. This requires some additional skills and motivations which can be referred to as shared or joint intentionality in the sense that the agent of the intentions and actions is the plural subject 'we'.

This capacity to collaborate—which has allowed humans to develop languages, sophisticated cultural systems, social norms, and so forth—relies, in turn, on other mechanisms such as joint attention, imitation, and "mind reading," all mechanisms that support the process of envisaging another person's point of view.[1]

There have been many studies in developmental psychology to capture the onset of children's capacity to "put themselves in other people's shoes," to envisage that others have different perspectives and thoughts, to "read other people's minds." Many experimental studies support the view that joint attention facilitates language development in our species (e.g., Baldwin and Moses 1996; Butterworth 2003). Developmental psychologists such as George Butterworth have extensively investigated the role of pointing and joint attention in an infant's initial stages of language acquisition and showed that this behaviour precedes and accompanies the onset of linguistic skills.

As individuals coordinate their actions with one another in collaborative activities with agent-neutral roles, they also coordinate their attention to things relevant to their joint goal—so-called joint attention (Bakeman and Adamson 1984; Moore and Dunham 1995). Children thus monitor the adult and her attention, who is, of course, monitoring them and their attention. No one is certain how best to characterise this potentially infinite loop of mutual monitoring (called "recursive mindreading" by Tomasello, 2008), but it seems to be part of infants' experience, in some nascent form, from before the first birthday. In addition to this shared attention on things, participants in these interactions each have their own perspective on things as well. Indeed, Moll and Tomasello (2007) argue that the whole notion of perspective depends on us first having a joint attentional focus, such as a topic, that we may then view differently—otherwise we just see completely different things. This dual-level attentional structure—shared focus of attention at a higher level, differentiated into perspectives at a lower level—is, of course, directly parallel to the dual-level intentional structure of the collaborative activity itself—a joint goal with individual roles—and ultimately derives from it (Tomasello 2011).

In understanding the importance of shared intentionality, or "we intentionality" as it is called by Gallotti and Frith (2013), and joint attention it is significant that it is a process specific to human infants. Tomasello describes the behaviour of apes, who also point and ensure in advance that the recipient is looking in the right direction. The ape's pointing, however, is generally an imperative request for something and never a declarative gesture aimed simply at informing or pointing something out.

Human infants, however, from the age of twelve months, start to point to objects not because they want them but because they seem to want to share a point of interest—they seem motivated to share psychological states. Infants start by following what others draw to their attention, then they learn to draw other people's attention to things that interest them. Eventually children learn to internalise language and symbols and become able to read the communicative intentions of others as embodied in their symbolic behaviour and to culturally learn those symbolic behaviours themselves. Language gradually becomes a way to draw attention to objects and situations that are more and more distant and to express more complex intentions.

Digital Communication and Attention

In essence, according to Tomasello, language is about sharing and directing attention. If he is right in his assertion, the essential connection between digital communication channels and attention is obvious. It is true that every time we receive a message and attend to it, we are being drawn into the attention of another. It is true also that when we craft an e-mail, a text, or a post, we are attempting to draw someone's attention into our space of intentions.

Our capacity for attracting others into our "attention 'space" through language and multiple channels is highly developed and sophisticated, and the fact that we are able to do this even in very short texts of less than 150 characters is a definite proof of this. The calls and messages that we create are effective precisely because they rely on our extraordinary capacity to "mind read," to understand other people's intentions, to see things from their perspective and make sense of messages that are frequently ridiculously underspecified. That is why a four-word text message can make us happy, aroused, or upset or a short post on Twitter can make us judge a person as trustworthy or as someone we would never want to date. This joining into other people's attention comes so naturally to us that we can do it even with fictional characters.

At the light of all the findings on language acquisition, it is possible that attention is only social and rarely just an individual process (Broadbent and Lobet-Maris 2014). However, in the digital world the most frequent metaphor used to refer to attention is an economic one: attention is a resource. An individual cognitive entity will have a limited "resource" that can be spent on something to the detriment of something else. Managing this limited resource is, therefore, a skill that is learnt and cultivated. People even speak of the "attention economy" to refer to the efforts deployed to make people "spend" their attention on one or another message. This individualistic metaphor presumes that attention is somehow linked to an act of volition (either conscious or not), and most cognitive theories have had to invoke a second order function that somehow controls attention, directing it by filtering out all distracting or competing sources of disturbance. Tomasello's (2008) account (in line with most of the recent studies in the field of language) paints a very different picture in which individuals are primarily social in their attention processes, moving in and out of shared goals and joint attentional states.

If we are always somehow collaborating, joining attention with others—either in copresence or through language, even at great temporal or spatial distance—the competition is not for our attention but shared goals. Who we join attention with and about what becomes the crucial issue, and the vast literature on how some social groups create a dominant discourse or are able to set the agenda becomes highly relevant.

If we put all this in the context of the phenomena I have described in previous chapters—people contacting their loved ones from their workplace, for instance—we can better understand the tensions that can arise. On the one hand, we have built complex social environments, institutions such as schools and administrations, that are construed to ensure systems of joint attention of shared goals and collaboration. On the other, we have recently flooded our environment with tools that draw people into other spaces of attention. McCullogh (2013) talks of "attention commons" to describe the spaces that have been constructed to share attention.

Traditionally, social environments have been organised to draw people together in the accomplishment of common goals and have relied on a variety of techniques for attention management such as artefacts to share intentions—the shape of buildings, the design of hierarchies, and roles and specific norms of behaviour. If we analyse a classroom, for instance, everything in its design is constructed to focus the pupils' attention: the layout of the room, which points to the place where the teacher sits, thus reinforcing the position of the teacher as a figure with a certain status and power; the posters on the walls, displaying the topic being studied; the shared artefacts such as syllabus and books; the establishment norms that regulate the pupils' behaviour. The whole group of students is, in theory, united with the teacher in a common effort of joint attention towards a topic or activity. Similar set-ups, in terms of both architecture and authority, are present in churches, boardrooms, and assemblies of all types. An office, a company, or a public administration are all highly complex social structures that also, at least in theory, are oriented towards achieving shared goals through highly elaborate collaborative activities. This is one reason why, at regular intervals, companies redefine their mission statements to remind employees of the joint purpose that explains and justifies the subtasks each person is accomplishing.

With such a deployment of effort to maintain the sense of joint attention, it is no wonder that the displacement of focus elicited by personal communication is seen as a challenge. Suddenly a child, a spouse, or a friend can penetrate, with a simple call, into the highly structured environment to demand—and obtain—immediate attention from any of the individuals who are part of the group. It is exactly because we know that people are so good at being drawn into distant attentional spaces with minimal cues that these interruptions are seen as pernicious.

So how do we resolve this tension?

The solution, in my view, is to be found in the theories of communication I have just discussed. These see us as sophisticated collaborators, adept at joining into others' attention. If it is indeed the case that our cognition is uniquely social, then what we are witnessing at the moment—this battle for attention—is simply a learning phase. As illustrated in Chapter 4, organisations that are ready to trust and negotiate seem to do quite well in keeping their members motivated and focused. People can move in and out of communicational settings with different objectives, switching between attention spaces and maintaining different shared goals. The working parents who tell us that they feel relieved when they know everything at home is fine or the couples who keep in touch during the day (as described in Chapter 1) are not questioning the relevance of their work objectives but rather simply juggling between spaces of shared intentionality.

What seems more challenging, however, is the ability to self-govern and self-direct our attention. If we are good at joining in and communicating, we are less skilful at maintaining attention alone, at sustaining extended periods of self-directed focus. Many workplaces' organisational and technological transformations in the last forty years have led to the fragmentation of tasks and more interaction with machines and processes than with people. Within these environments competing with personal communication devices is a lost battle. Certain activities, whether operating a machine or checking forms, often fail to keep the worker's attention. In these contexts a personal communication device offers far more compelling potential for a collaborative joining into attention. What personal channels reveal is that some institutions may have been stretching our capacity for isolated activity too far.

Notes

Chapter 2

1. Farmville is a virtual game created by Zinga in which users manage a virtual farm by ploughing land, planting, growing, and harvesting virtual crops, harvesting trees and bushes, and raising livestock. In 2010 it had 62 million users. It can be accessed through Facebook or on iPhones and Android smartphones.

2. As will be discussed in the following chapter, one reading of the intensity of small group exchanges can be found in new forms of solidarity and mutualism. The economic crisis in some regions of Europe, for instance, which is reducing not only access to jobs but also their duration and conditions, is transforming young people's sense of agency. New strategies for coping with contingency are emerging that, depending on factors such as ethnicity, socioeconomic origin, geographical area, and so on, seem to be centred on reconfiguring young people's relations to classical institutions. In order to tackle discontinuities in life patterns, people take on new forms of distribution of risk. These take the form of stronger intergenerational links, a strengthening of core social ties, numerous forms of collaboration (in consumption, housing, crowdsourcing, and crowdfunding), and the rise of networks of self-employment. Young people live longer with parents, live in hostels, share houses and workplaces, work from home, use crowdsourcing resources for funding for sharing goods, and create strong personal networks of support. All these phenomena can be read either as forms of withdrawal into the private sphere or as ways to reconfigure the social sphere in order to socially manage new economic and professional demands. These reconfigurations can provide a glimpse of the directions and solutions that can be expected in the near future.

Chapter 3

1. A study of the surfing behaviours of employees in forty-four companies in France by Olfeo showed a different set of activities according to the hour of day: 9 to 10 a.m. for news, e-mail, and social networks; 11 a.m. to 2:00 p.m. for entertainment sites such as videos, music, and games; and 5 to 6:00 p.m. for services such as transportation, traffic, and weather. (Olfeo 2012).

2. We cannot avoid a parallel with the forms of control described by Foucault in his Surveiller et Punir. The presence of video cameras on factory floors have a very clear panopticon undertone.

3. Using the mobile phone in moments of distress or extreme anxiety is sometimes

dramatically reported by the news, as during the Mumbai attacks in 2008. Many of the hostages kept in touch through SMS, e-mail, and calls during the long hours they were held captive. Recordings of calls made to relatives during the September 11 attacks either from the planes or from the buildings are also being used as evidence. On a less tragic scale, most people interviewed report of cases in which their mobiles were essential to solve a sticky situation: a car breakdown, a loss of keys, or being lost somewhere. The equation between a perception of safety and access to communication technology seems rather strong and has to be reckoned with in explanations regarding mediated communication.

Chapter 4

1. In August 2010 CNN reported that President Barack Obama signed into law a prohibition on cell phone use by prisoners. The law prohibits the use or possession of mobile phones and wireless devices, and calls for up to a year in prison for anyone found guilty of trying to smuggle one to an inmate. The Federal Bureau of Prisons confiscated more than 2,600 cell phones from minimum security facilities and nearly 600 from secure federal institutions last year.
2. See Dares Direction de l'animation de la recherche, des etudes et des statistiques 2010.
3. Just to confirm our argument on the differences in access to private communication as a function of the social status of their job, the 2009 PEW research by Madden and Jones indicated that

> The same groups who report the most frequent checking of work-related email at work are also the most likely to report frequent tending to their personal accounts while at work: higher-income workers and those tethered to a desk. Fully 66% of those in jobs earning $75,000 per year or more say they check their personal accounts at work, compared with just 45% of those in jobs earning less than $30,000 annually. Likewise, those in professional, managerial and clerical positions are more tuned into their personal e-mail at work when compared with those working in the service industry, skilled and semi-skilled jobs.

4. See Berret 2008.

Chapter 5

1. The Union of Transport has a document regarding crew and personnel tiredness and its link to accidents. It emerges that, especially in freight trains, crews can self-regulate the amount of hours they work. It is, therefore, not uncommon, they say, to have train engineers who have slept six hours in the last forty-eight hours or engineers who have not slept at all for twenty-seven hours. Furthermore, crews can be called at the last minute to drive a train, and it is common for freight crews to have fifteen different hours of start of duty in fifteen days, ranging from 2 a.m. to 7 p.m. Falling asleep in the cab and thus missing red lights has been a reported cause of many accidents. Train companies have reduced their personnel over the years, and as a result, the existing crews are asked to step in at all times.

However, unions are defending this type of work organisation because it allows personnel to work more hours and, thus, earn more.

Chapter 6

1. In fact, most modern theories of communication seem to rely on the presupposition of a collaborative intent that will help to disambiguate the inherent underdeterminacy of utterances.(Sperber and Wilson 1995)

References

Aaltonen, A., and Kallinikos, J. 2013. "Coordination and Learning in Wikipedia: Revisiting the Dynamics of Exploitation and Exploration." *Research in the Sociology of Organizations* 37: 161–192.

Ariès, P. 1977. The Family and the City. *Daedalus* (106) 2: 227–235.

Ariès, P., and G. Duby. 1985–1986. *Histoire de la vie privée*. Paris: Seuil.

Bakeman, R., and L. B. Adamson. 1984. "Coordinating Attention to People and Objects in Mother-Infant and Peer-Infant Interaction." *Child Development* (55): 1278–1289.

Baldwin, D. A., and L. J. Moses. 1996. "The Ontogeny of Social Information Gathering." *Child Development* (67): 1915–1939.

Baumann, Z. 2006. *Liquid Times: Living in an Age of Uncertainty*. Cambridge: Polity.

Baym, N. 2010. *Personal Connections in the Digital Age*. London: PolityPress.

Bernstein, M., E. Bakshy, M. Burke, and B. Karrer. 2013. "Quantifying the Invisible Audience," ACM Conference on Human Factors in Computing Systems (CHI), April 29, 2013.

Berret, P. 2008. "Diffusion et utilization de TIC en France et en Eu-rope." Culture chiffres (2): www2.culture.gouv.fr/culture/deps/2008/pdf/ cchiffres08_2.pdf.

Boase, J., and B. Wellman. 2006. "Personal Relationships: On and Off the Internet." In *The Cambridge Handbook of Personal Relationships,* edited by A. Vangelisti and E. D. Perlman, Cambridge: Cambridge University Press: 709–723.

Boellstorff, T. 2008. *Coming of Age in Second Life: An Anthropologist Explores the Virtually Human*. Princeton, NJ: Princeton University Press.

Bowlby, J. 1979. *The making and breaking of affectionate bonds*. London: Tavistock Publications Ltd.

Bowlby, J., M. Ainsworth, M. Boston, and D. Rosenbluth. 1956. "The Effects of Mother-Child Separation: A Follow-up Study." *British Journal of Medical Psychology* (29): 211–247.

Boyd, D. 2007a. "Viewing American Class Divisions Through Facebook and MySpace." Apophenia Blog Essay. www.danah.org/papers/essays/ClassDivisions.html.

Boyd, D. 2007b. "Why Youth (Heart) Social Network Sites: The Role of Networked Publics in Teenage Social Life." In *Youth, Identity, and Digital Media Volume,* edited by D. Buckingham, pp. 119–142. Cambridge, MA: MIT Press.

Boyd, D. 2014. *It's Complicated: The Social Life of Networked Teenagers*. New Haven, CT: Yale University Press.

Broadbent, S. 2001. "Evolution des usages de l'Internet." In *Comprendre les usages de l'Internet,* edited by E. Guichard, pp. 156–165. Paris, Presses de l'ecole Normale Supérieure.

Broadbent, S. 2012. "Approaches to Personal Communication." In *Digital Anthropology,* edited by H. Horst and D. Miller: 95–112.

Broadbent, S., and V. Bauwens. 2008. "Understanding Convergence." *Interactions* (15) 1: 29–37.

Broadbent, S., and F. Cara. 2003. "The New Architectures of Information." In *Texte-e: Le texte à l'heure d'Internet,* edited by G. Origgi and N. Arikha, pp. 197–215. Paris, BPI.

Broadbent, S., and C. Lobet-Maris. 2014. "Towards a Grey Ecology." In *The Onlife Manifesto: Being Human in a Hyperconnected Era,* edited by L. Floridi: 111–124. Dordrecht: Springer.

Brugière, A., and A. Rivière. 2010. *Bien vieillir grâce au numérique*. Limoges: FYP éditions.

Brynjolfsson, E., and A. McAfee. 2014. *The Second Machine Age: Work, Progress, and Prosperity in a Time of Brilliant Technologies*. New York: W. W. Norton Company.

Bryson, V. 2007. *Gender and the Politics of Time*. Bristol: Polity Press.

Bunting, M. 2004. *Willing Slaves: How the Overwork Culture Is Running Our Lives*. London: Harper Collins.

Burke, M., and R. Kraut. 2014. "Growing Closer on Facebook: Changes in Tie Strength Through Site Use." ACM CHI 2014: Conference on Human Factors in Computing Systems, New York: ACM.

Burke, M., R. Kraut, and C. Marlow. 2013. "Using Facebook After Losing a Job: Differential Benefits of Strong and Weak Ties." ACM CSCW 2013: Conference on Computer-Supported Cooperative Work, pp. 1419–1430. New York: ACM.

Butterworth, G. 2003. "Pointing Is the Royal Road to Language for Babies." In *Pointing: Where Language, Culture, and Cognition Meet,* edited by S. Kita, pp. 9–33. Mahwah, NJ: Lawrence Erlbaum.

Castells, M. 2006. *Mobile Communication and Society: A Global Perspective*. Cambridge, MA: MIT Press.

Coleman, E. G. 2012. *Coding Freedom: The Ethics and Aesthetics of Hacking*. Princeton, NJ: Princeton University Press.

Comision de investigation de accidentes ferroviarios. 2013. "Informe final investigation de l'accidente grave 0054/13 occurido el 24/07/2013." www.fomento.gob.es/NR/rdonlyres/0ADE7F17-84BB-4CBD-9451-C750EDE06170/125127/IF240713200514CIAF.pdf.

Dares. (Direction de l'animation de la recherche, des études et des statistiques) 2010. "Activité et conditions d'emploi de la main-d'oeuvre au 3e trimestre 2010." *Dares indicateurs* (76, November).

Derber, C. 2000. *The Pursuit of Attention: Power and Ego in Everyday Life,* 2nd ed. Oxford: Oxford University Press.

Donner, J. 2006. "The Use of Mobile Phones by Microentrepreneurs in Kigali, Rwanda: Changes to Social and Business Networks." *Information Technologies and International Development* (3) 2: 3–19.

Donner, J. 2007. "The Rules of Beeping: Exchanging Messages via Intentional 'Missed Calls' on Mobile Phones." *Journal of Computer-Mediated Communication* (13) 1: 1–22.

Donner, J. 2009. "Blurring Livelihoods and Lives the Social Uses of Mobile Phones and Socioeconomic Development." *Innovations* (4) 1: 91–101.

Duggan, M., Lenhart, A., Lampe, C., Ellison, N.B. July 2015. "Parents and Social Media." Pew Research Center, www.pewinternet.org/2015/07/16/parents-and-social-media/.

Dutton, W. H., and F. Nainoa. 2002. "Say Goodbye ... Let's Roll: The Social Dynamics of Wireless Networks on September 11." *Prometheus* (20) 3: 237–245.

Elias, N. 1939. *Über den Prozess der Zivilisation: Soziogenetische und psychogenetische Untersuchungen,* vol. 1: *Wandlungen des Verhaltens in den weltlichen Oberschichten des Aben-dlandes.* Basel: Haus zum falken.

Ellwood-Clayton, B. 2003. "Virtual Strangers: Young Love and Texting in the Filipino Archipelago of Cyberspace." In *Mobile Democracy: Essays on Society, Self and Politics,* edited by K. Nyiri: 225–239. Wien: Passagen Verlag.

Ellwood-Clayton, B. 2005. "Desire and Loathing in the Cyber Philippines." In *The Inside Text: Social, Cultural and Design Perspectives on SMS,* edited by R. Harper, L. Palen, and A. Taylor: 195–219. Dordrecht: Springer.

Facebook Data Team. 2009, March 9. "Maintained Relationships on Facebook." www.facebook.com/data?v=app_4949752878#!/notes/facebook-data-team/maintained-relationships-on-facebook/55257228858.

Finnegan, G., R. Howarth, and P. Richardson. 2004. "The Challenges of Growing Small Businesses: Insights from Women Entrepreneurs in Africa." Genève, International Labour Organization, SeeD Working Paper, n. 47.

Fischer, C. 1992. *America Calling: A Social History of the Telephone.* Berkeley: University of California Press.

Foucault, M. 1975. *Surveiller et punir.* Paris: Gallimard.

Furedi, F. 2002. *Paranoid Parenting: Why Ignoring the Experts May Be the Best for Your Child.* Chicago, IL: Chicago Review Press.

Furedi, F. 2009. "Precautionary Culture and the Rise of Possibilistic Risk Assessment." *Erasmus Law Review* (2) 2: 197–220.

Gallotti, M., and C. D. Frith. 2013. "Social Cognition in the We-Mode." *Trends in Cognitive Sciences* (17): 160–165.

Gershon, I. 2010. *The Breakup 2.0: Disconnecting Over New Media.* Ithaca, NY: Cornell University Press.

Goffman, E. 1959. *The Presentation of Self in Everyday Life.* Edinburgh: Anchor Books.

Gournay, C. de. 2002. "Pretense of Intimacy in France." In *Perpetual Contact,* edited by J. E. Katz and E. M. Aakhuse, 193–205. Cambridge: Cambridge University Press.

Gournay, C. de, and Z. Smoreda. 2001. "Technologies de communication et relations de proximité." *Annales de la recherche urbaine* (90): 67–76.

Granovetter, M. 1973. "The Strength of Weak Ties". *American Journal of Sociology* 78: 1360–1380.

Grice, P. 1989. *Studies in the Way of Words.* Cambridge, MA: Harvard University Press.

Habermas, J. 1962. *Strukturwandel der Öffentlichkeit.* Neuwied-Berlin: Luchterhand.

Hall, J. A., and N. K. Baym. 2012. "Calling and Texting (Too Much): Mobile Maintenance Expectations (Over) Dependence, Entrapment, and Friendship Satisfaction." *New Media and Society* (14) 2: 316–331.

Hampton, K., L. Goulet, J. L. Eun, and L. Rainie. 2009. "Social Isolation and New Technology: How the Internet and Mobile Phones Impact Americans' Social Networks." Pew Internet and American Life Project. Washington, DC.

Hand, M. 2012. *Ubiquitous Photography.* London: Polity.

Hoc, J. M. 2001. "Towards a Cognitive Approach to Human-Machine Cooperation in Dynamic Situations." *International Journal of Human Computer Studies* (54) 4: 509–540.

Hollnagel, E., D. Woods, and N. Leveson. 2006. *Resilience Engineering in Practice: A Guidebook.* Farnham: Ashgate.

Horst, H., and D. Miller. 2006. *The Cell Phone: An Anthropology of Communication*. London: Berg.

Horst, H., and D. Miller. 2012. *Digital Anthropology*. London: Berg.

Hutchins, E. 1995. *Cognition in the Wild*. Cambridge, MA: MIT Press.

Ito, M. 2010. *Hanging Out, Messing Around, and Geeking Out: Kids Living and Learning with New Media*. Cambridge, MA: MIT Press.

Ito, M., and D. Okabe. 2005. "Contextualizing Japanese Youth and Mobile Messaging." In *The Inside Text: Social, Cultural and Design Perspectives on SMS*, edited by R. Harper, L. Palen, and E. A. Taylor Dordrecht: Springer: 127–145.

Ito, M., D. Okabe, and M. Matsuda. 2005. *Personal, Portable, Pedestrian, Mobile Phones in Japanese Life*. Cambridge, MA: MIT Press.

Kahneman, D. 1979. *Attention and Effort*. Englewood Cliffs, NJ: Prentice-Hall.

Kallinikos, J. 2011. *Governing Through Technology: Information Artefacts and Social Practice: Technology, Work and Globalization*. Basingstoke, UK: PalgraveMacmillan.

Kallinikos, J., A. Aaltonen, and A. Marton. 2013. "The Ambivalent Ontology of Digital Artifacts." *MIS Quarterly* 37: 357–370.

Kaplan, D. 2010. *Informatique, libertés, identités*. Limoges: FYP Éditions.

Kasesniemi, E. 2003. *Mobile Messages*. Tampere, Finland: Tampere University Press.

Kelty, C. 2008. *Two Bits: The Cultural Significance of Free Software*. Durham, NC: Duke University Press.

Kim, K.-H., and Y. Han. 2007. "Cying for Me, Cying for Us: Relational Dialectics in a Korean Social Network Site." *Journal of Computer-Mediated Communication* (13) 1, art. 15: http://jcmc.indiana.edu/vol13/issue1/kim.yun.html.

Klein, N. 2007. *The Shock Doctrine. The Rise of Disaster Capitalism*. New York, Metropolitan Books.

Lee, D. 2009. "The Impact of Mobile Phones on the Status of Women in India." PhD Diss., Stanford University. www.mobileactive.org/files/file_uploads/mobilePhonesandWom-eninindia.pdf.

Lenhart, A. 2015. "Teens, Social Media, and Technology Overview 2015." Pew Research Center. www.pewinternet.org/2015/04/09/teens-social-media-technology-2015.

Ling, R. S. 2004. *The Mobile Connection: The Cell Phone's Impact on Society*. San Francisco, CA: Morgan Kaufman.

Ling, R. S. 2008. *New Tech, New Ties: How Mobile Communication Is Reshaping Social Cohesion*. Cambridge, MA: MIT Press.

Livingstone, S., G. Mascheroni, and M. F. Murru. 2014. "Social Networking Among European Children: New Findings on Privacy, Identity and Connection." In *Identité(s) Numérique(s). Les Essentials d'Hermès*, edited by Wolton, Dominique. Paris: CNRS Editions.

Macfarlane, A. 2002. *The Making of the Modern World: Visions from the West and from the East*. London: Palgrave Macmillan.

Madden, M., and S. Jones. 2008, September 24. "Networked Workers." Pew Research Center, Pew Internet and American Life Project. www.pewinternet.org/2008/09/24/networked-workers.

Madden, M., and L. Raine. 2010. "Adults and Cell Phone Distractions." Pew Research Center. www.pewinternet.org/2010/06/18/adults-and-cell-phone-distractions.

Madianou, M., and D. Miller. 2012. *Migration and New Media: Transnational Families and Polymedia*. London: Routledge.

Marchbank, J. 2004. *Women, Power and Policy: Comparative Studies of Childcare*. New York: Routledge.

Mauss, M. 1954. *The Gift: Forms and Functions of Exchange in Archaic Societies*. London: Cohen and West.

McCullogh, M. 2013. *Ambient Commons: Attention in the Age of Embodied Information*. Cambridge, MA: MIT Press.

Mikulincer, M., and P.R. Shaver. 2007. *Attachment in Adulthood: Structures, Dynamics and Change*. New York: Guilford Press.

Miller, D. 2011. *Tales from Facebook*. Cambridge: Polity Press.

Moll, H., and M. Tomasello. 2007. "Co-operation and Human Cognition: The Vygotskian Intelligence Hypothesis," *Philosophical Transactions of the Royal Society* (362): 639–648.

Moore, C., and P. J. Dunham, eds. 1995. *Joint Attention: Its Origins and Role in Development*. Hillsdale, NJ: Erlbaum.

Naaman, M., J. Boase, and C. Lai. 2010. "Is It Really About Me? Message Content in Social Awareness Streams." In *Proceedings of the 2010 ACM Conference on Computer-Supported Cooperative Work*, pp. 189–192. New York: ACM.

Nielsen. 2014. "Digital Consumer Report." www.nielsen.com/content/dam/corporate/us/en/reports-downloads/2014%20Reports/the-digital-consumer-report-feb-2014.pdf.

Nippert-Eng, C. E. 1996. *Home and Work: Negotiating Boundaries Through Everyday Life*. Chicago: University of Chicago Press.

NTSB (National Transportation Safety Board). 2010a. "Collision of Metrolink Train 111 with Union Pacific Train LOF65–12, Chatsworth, California, September 12, 2008." Accident report ntSb/rar-10/01. Washington, DC.

NTSB (National Transportation Safety Board). 2010b. "Midair Collision Over Hudson River, Piper PA-32R-300, N71MC and Eurocopter AS350BA, N401LH Near Hoboken, New Jersey, August 8, 2009." Accident Report NTSB/aar-10/05. Washington, DC.

Oatley, K., D. Keltner, and J. Jenkins. 1996. *Understanding Emotions*. Oxford: Blackwell.

OFCOM (Office of Communications). 2014. "Communications Market Report, August 2014." http://stakeholders.ofcom.org.uk/market-data-research/market-data/communications-market-reports/cmr14.

Olfeo. 2012. *La realité de l'utilisation d'internet au bureau*. Etude OLFEO 2012.

ONS (Office National Statistics). 2014. "Self Employed Workers in the UK—2014." www.ons.gov.uk/ons/dcp171776_374941.pdf.

Park, Y., G. Heo, and R. Lee. 2008. "Cyworld Is My World: Korean Adult Experiences in an Online Community for Learning." *International Journal Web Based Communities* (4) 1: 33–51.

Perrow, C. 1984. *Normal Accidents: Living with High-Risk Technology*. Princeton, NJ: Princeton University Press.

Pryanka, C. 2010. "Mobile Phone Usage by Young Adults in India: A Case Study." PhD diss., University of Maryland.

Putnam, R. D. 2000. *Bowling Alone: The Collapse and Revival of American Community*. New York: Simon and Schuster.

Rainie, L., Wellman, B. 2012. *Networked: The New Social Operating System*. Boston Mass.: MIT Press.

Ramirez Angel, M. 2016. "Colombia's Digital Orphans and Childless Parents: maintaining sentiments across national borders." PhD diss., University College London.

Ramirez Angel, M., and S. Broadbent. 2015 (forthcoming). "An Ethnography of Patient Support Groups." London: Nesta.

Ryan, D. 2006. "Getting the Word Out: Notes on Social Organization of Notification." *Sociological Theory* (24) 3: 228–254.

Schnorf, S. 2008. *Diffusion in sozialen: Netzwerken der Mobilkommunikation*. Zürich: UVK.

Scholz, T. 2013. *Digital Labour: The Internet as Playground and Factory*. New York: Rutledge.

Sennett, R. 1977. *The Fall of Public Man*. Cambridge: Cambridge University Press.

Sennett, R. 1998. *The Corrosion of Character: The Personal Consequences of Work in the New Capitalism*. New York: Norton.

Simmel, G. 1971. *On Individuality and Social Forms: Selected Writings*. Chicago: University of Chicago Press.

Spencer, E., and R. Pahl. 2006. *Rethinking Friendship: Hidden Solidarities Today*. Princeton, NJ: Princeton University Press.

Sperber, D., and D. Wilson. 1995. *Relevance: Communication and Cognition*, 2nd ed. Oxford: Blackwell.

Standing, G. 2011. *The Precariat: The New Dangerous Class*. London: Bloomsbury.

Statista. 2014. "Statistics and Facts About Whatsapp." www.statista.com/topics/2018/whatsapp.

Strayer, D. L. 2006. "A Comparison of the Cell Phone Driver and the Drunk Driver." *Human Factors* (48) 2: 381–391.

Takahashi, T. 2010. "MySpace or Mixi?: Japanese Engagement with SNS (Social Networking Sites) in the Global Age." *New Media Society* (12): 453–475.

Takhteyev, Y., S. Gruzd, and B. Wellman. 2012. "Geography of Twitter Networks." *Social Networks* (34) 1: 73–81.

Tomasello, M. 2008. *Origins of Human Communication*. Cambridge, MA: MIT Press.

Tomasello, M. 2011. "Human Culture in Evolutionary Perspective." In *Advances in Culture and Psychology*, edited by M. Gelfand, Chi-Yue Chiu, and Ying-Yi Hong, Oxford: Oxford University Press: 5–53.

Tomasello, M., M. Carpenter, J. Call, T. Behne, and H. Moll. 2005. "Understanding and Sharing Intentions: The Origins of Cultural Cognition." *Behavioral and Brain Sciences* (28): 675–691.

Tomkins, S. 1962. *Affect Imagery Consciousness*, vol. 1: *The Positive Affects*. London: Tavistock.

Turkle, S. 2011. *Alone Together: Why We Expect More from Technology and Less from Each Other.* New York: Basic Books.

van Deursen, A., J. van Dijk, 2011. "Internet Skills and the Digital Divide." *New Media and Society* (13): 893–911.

Venkatraman, S. 2015. "Women Entrepreneurs and WhatsApp." The Global Social Media Impact Study, UCL Department of Anthropology. http://blogs.ucl.ac.uk/global-social-media/2015/04/17/women-entrepreneurs-whatsapp.

Venkatraman, S. 2016. *Social Media in South India*. London. University College London Press.

Vygotskii, L. S. 1978. Mind in Society: The Development of Higher Psychological Processes. Cambridge, MA: Harvard University Press.

Wajcman, J., M. Bittman, J. E. Brown. 2008. "Families without Borders: Mobile Phone Connectedness and Work-Home Divisions." *Sociology* 42 (4): pp. 635–652.

Wallis, C. 2008. "Techno-Mobility and Translocal Migration: Mobile Phone Use Among Female Migrant Workers in Beijing." In *Gender Digital Divide*, edited by M. I. Sri-Nivasan and V. V. Ramani, Hyderabad: Icfai University Press: 196–216.

Wallis, C. 2013. *Techno-Mobility in China: Young Migrant Women and Mobile Phones*. New York: New York University Press.

Wang, X. 2015. "Archive for the 'China (South)' Category." The Global Social Media Impact Study, UCL Department of Anthropology, http://blogs.ucl.ac.uk/global-social-media/category/fieldsites/china-south.

Wellman, B. 1979. "The Community Question: The Intimate Networks of East Yorkers." *American Journal of Sociology* (84): 1201–1231.

Wellman, B., K. Hampton, A. Quan-Haase, and J. Witte. 2001. "Does the Internet Increase, Decrease, or Supplement Social Capital?: Social Networks, Participation, and Community Commitment." *American Behavioral Scientist* (45) 3: 437–456.

Wellman, B., and C. Haythornthwaite, eds. 2002. *The Internet in Everyday Life*. Oxford: Blackwell.

Wellman, B., A. Smith, A. Wells, and T. Kennedy. 2008. "Networked Families." Pew Research Center, Pew Internet and American Life Project. www.pewinternet.org/2008/10/19/networked-families.

Winnicott, D. W. 1965. *The Maturational Processes and the Facilitating Environment*. London: Karnac Books.

Zaretsky, E. 1976. *Capitalism, the Family and Personal Life*. New York: Harper & Row.

Index

About the Author

Stefana Broadbent earned her PhD in cognitive science from the University of Edinburgh and contributed to *The Onlife Manifesto* (Springer, 2015) and *Digital Anthropology* (Bloomsbury Academic, 2012). Editions of *Intimacy at Work* have been published in French (*L'Intimité au Travail*, FYP Editions, 2011) and Italian (*Internet lavoro e vita private*, Il Mulino, 2013). For the last twenty years Broadbent has studied the social, cultural, and cognitive aspects involved in the use of technology at work and at home. She is currently Head of Collective Intelligence at Nesta, an independent charitable organization in the UK, where she conducts research into how networked groups find new ways to collaborate with one another. Previously she was a lecturer in digital anthropology in the Department of Anthropology at University College in London.

Franklin Pierce University
00209548